Periodic Table of Elements

Atomic Number → | 1 **H** Hydrogen 1.008 | ← Symbol

Name → | | ← Atomic Weight

1 IA	2 IIA	3 IIIB	4 IVB	5 VB	6 VIB	7 VIIB	8 VIIIB	9 VIIIB
1 **H** Hydrogen 1.008								
3 **Li** Lithium 6.94	4 **Be** Beryllium 9.0121831							
11 **Na** Sodium 22.98976928	12 **Mg** Magnesium 24.305							
19 **K** Potassium 39.0983	20 **Ca** Calcium 40.078	21 **Sc** Scandium 44.955908	22 **Ti** Titanium 47.867	23 **V** Vanadium 50.9415	24 **Cr** Chromium 51.9961	25 **Mn** Manganese 54.938044	26 **Fe** Iron 55.845	27 **Co** Cobalt 58.933194
37 **Rb** Rubidium 85.4678	38 **Sr** Strontium 87.62	39 **Y** Yttrium 88.90584	40 **Zr** Zirconium 91.224	41 **Nb** Niobium 92.90637	42 **Mo** Molybdenum 95.95	43 **Tc** Technetium (98)	44 **Ru** Ruthenium 101.07	45 **Rh** Rhodium 102.90550
55 **Cs** Caesium 132.90545196	56 **Ba** Barium 137.327	57 - 71 Lanthanoids	72 **Hf** Hafnium 178.49	73 **Ta** Tantalum 180.94788	74 **W** Tungsten 183.84	75 **Re** Rhenium 186.207	76 **Os** Osmium 190.23	77 **Ir** Iridium 192.217
87 **Fr** Francium (223)	88 **Ra** Radium (226)	89 - 103 Actinoids	104 **Rf** Rutherfordium (267)	105 **Db** Dubnium (268)	106 **Sg** Seaborgium (269)	107 **Bh** Bohrium (270)	108 **Hs** Hassium (269)	109 **Mt** Meitnerium (278)

57 **La** Lanthanum 138.90547	58 **Ce** Cerium 140.116	59 **Pr** Praseodymium 140.90766	60 **Nd** Neodymium 144.242	61 **Pm** Promethium (145)	62 **Sm** Samarium 150.36	63 **Eu** Europium 151.964	64 **Gd** Gadolinium 157.25
89 **Ac** Actinium (227)	90 **Th** Thorium 232.0377	91 **Pa** Protactinium 231.03588	92 **U** Uranium 238.02891	93 **Np** Neptunium (237)	94 **Pu** Plutonium (244)	95 **Am** Americium (243)	96 **Cm** Curium (247)

10 VIIIB	11 IB	12 IIB	13 IIIA	14 IVA	15 VA	16 VIA	17 VIIA	18 VIIIA
								2 **He** Helium 4.002602
			5 **B** Boron 10.81	6 **C** Carbon 12.011	7 **N** Nitrogen 14.007	8 **O** Oxygen 15.999	9 **F** Fluorine 18.998403163	10 **Ne** Neon 20.1797
			13 **Al** Aluminium 26.9815385	14 **Si** Silicon 28.085	15 **P** Phosphorus 30.973761998	16 **S** Sulfur 32.06	17 **Cl** Chlorine 35.45	18 **Ar** Argon 39.948
28 **Ni** Nickel 58.6934	29 **Cu** Copper 63.546	30 **Zn** Zinc 65.38	31 **Ga** Gallium 69.723	32 **Ge** Germanium 72.630	33 **As** Arsenic 74.921595	34 **Se** Selenium 78.971	35 **Br** Bromine 79.904	36 **Kr** Krypton 83.798
46 **Pd** Palladium 106.42	47 **Ag** Silver 107.8682	48 **Cd** Cadmium 112.414	49 **In** Indium 114.818	50 **Sn** Tin 118.710	51 **Sb** Antimony 121.760	52 **Te** Tellurium 127.60	53 **I** Iodine 126.90447	54 **Xe** Xenon 131.293
78 **Pt** Platinum 195.084	79 **Au** Gold 196.966569	80 **Hg** Mercury 200.592	81 **Tl** Thallium 204.38	82 **Pb** Lead 207.2	83 **Bi** Bismuth 208.98040	84 **Po** Polonium (209)	85 **At** Astatine (210)	86 **Rn** Radon (222)
110 **Ds** Darmstadtium (281)	111 **Rg** Roentgenium (282)	112 **Cn** Copernicium (285)	113 **Nh** Nihonium (286)	114 **Fl** Flerovium (289)	115 **Mc** Moscovium (289)	116 **Lv** Livermorium (293)	117 **Ts** Tennessine (294)	118 **Og** Oganesson (294)

65 **Tb** Terbium 158.92535	66 **Dy** Dysprosium 162.500	67 **Ho** Holmium 164.93033	68 **Er** Erbium 167.259	69 **Tm** Thulium 168.93422	70 **Yb** Ytterbium 173.045	71 **Lu** Lutetium 174.9668
97 **Bk** Berkelium (247)	98 **Cf** Californium (251)	99 **Es** Einsteinium (252)	100 **Fm** Fermium (257)	101 **Md** Mendelevium (258)	102 **No** Nobelium (259)	103 **Lr** Lawrencium (266)

A NOTE ON THE AUTHOR

Kit Chapman is an award-winning science journalist and broadcaster. Initially qualifying as a pharmacist, Chapman began his career on medical journal *The Practitioner* before moving to *Chemist+Druggist*, the UK's leading magazine for pharmacists. After stints as campaign website manager for the British Medical Association and clinical editor for *The Pharmaceutical Journal*, Chapman was appointed comment editor for *Chemistry World*.

Chapman writes for the *Daily Telegraph* and *Sunday Telegraph*, and has appeared as an expert for the BBC and Sky News.

Also available in the Bloomsbury Sigma series:

Sex on Earth by Jules Howard
Spirals in Time by Helen Scales
A Is for Arsenic by Kathryn Harkup
Herding Hemingway's Cats by Kat Arney
Death on Earth by Jules Howard
The Tyrannosaur Chronicles by David Hone
Soccermatics by David Sumpter
Big Data by Timandra Harkness
Goldilocks and the Water Bears by Louisa Preston
Science and the City by Laurie Winkless
Bring Back the King by Helen Pilcher
Built on Bones by Brenna Hassett
The Planet Factory by Elizabeth Tasker
Wonders Beyond Numbers by Johnny Ball
I, Mammal by Liam Drew
Reinventing the Wheel by Bronwen and Francis Percival
Making the Monster by Kathryn Harkup
Catching Stardust by Natalie Starkey
Seeds of Science by Mark Lynas
Eye of the Shoal by Helen Scales
Nodding Off by Alice Gregory
The Science of Sin by Jack Lewis
The Edge of Memory by Patrick Nunn
Turned On by Kate Devlin
Borrowed Time by Sue Armstrong
We Need to Talk About Love by Laura Mucha
The Vinyl Frontier by Jonathan Scott
Clearing the Air by Tim Smedley
Genuine Fakes by Lydia Pyne
Grilled by Leah Garcés
The Contact Paradox by Keith Cooper
Life Changing by Helen Pilcher
Friendship by Lydia Denworth
Death by Shakespeare by Kathryn Harkup
Sway by Pragya Agarwal
Bad News by Rob Brotherton
Unfit for Purpose by Adam Hart
Mirror Thinking by Fiona Murden
Kindred by Rebecca Wragg Sykes
First Light by Emma Chapman

SUPERHEAVY

Making and Breaking the Periodic Table

Kit Chapman

BLOOMSBURY SIGMA
LONDON • OXFORD • NEW YORK • NEW DELHI • SYDNEY

BLOOMSBURY SIGMA
Bloomsbury Publishing Plc
50 Bedford Square, London, WC1B 3DP, UK
29 Earlsfort Terrace, Dublin 2, Ireland

Originally published in the United Kingdom in 2019
This edition published 2021

A catalogue record for this book is available from the British Library

Library of Congress Cataloguing-in-Publication data has been applied for

ISBN: PB: 978-1-4729-5392-6; eBook: 978-1-4729-5391-9

6 8 1 0 9 7 5

Typeset by Deanta Global Publishing Services, Chennai, India
Printed and bound in Great Britain by CPI Group (UK) Ltd, Croydon CR0 4YY

Bloomsbury Sigma, Book Forty-five

To find out more about our authors and books visit www.bloomsbury.com and sign up for our newsletters

Contents

Prologue

In January 2016 I picked a fight with a radio DJ. A few weeks earlier, I'd been listening to a podcast when the BBC's Simon Mayo was asked how many chemical elements there were. '118,' he replied on instinct.

I remember furrowing my brow. Despite being a science journalist, my knowledge of the freaky end of the periodic table – where the elements existed for fractions of a second and didn't seem to count – was poor. However, I was certain there were only 114: the elements up to 112, then 114 and 116 out on their own. I went home and checked.

The answer was 114. Mayo was wrong.

Then, the discoveries of four new elements were confirmed – all the way up to element 118. I didn't have much to do, so I hassled Mayo over Twitter about his prescience. 'The story has been around a long time!' he fired back. Annoyed at myself, I started brushing up on the new elements – if only so I could send more snippy messages to DJs in the future. A few hours later, I realised that I'd been missing out on one of the greatest untold sagas in science. For three generations, modern element discovery has seen heroes, villains, natural disasters, motor races, crash landings and giant particle cannons. It has given us nuclear power, atomic weapons, cancer treatments, smoke detectors and Kentucky Fried Chicken (really). It has united countries and driven Cold War rivals apart.

The superheavy elements – the elements from 104 and beyond – might only last for seconds, but that's what makes them so cool. When an atom of a superheavy element is created, it is probably the only atom of that element in existence *in the galaxy*. It's a science for pioneers and dreamers.

Superheavy is about more than the elements themselves. Too often, science is viewed as the purview of old white men. My hope is that, by reading this, you'll realise this is untrue. The superheavy element discoverers were a mix of ages, nationalities, ethnicities and genders. Too few people know that in 1952 a 28-year-old pilot, Jimmy Robinson, died on a mission that led to the discovery of two elements; that an African American, James Harris, was a key part of the team that discovered element 104; or that it was a woman, Darleane Hoffman, who led the discovery of the rarest natural element deposit ever found on Earth. Science doesn't care what you look like or where you come from.

This tale has been a global endeavour. To tell it, I've travelled to eight countries on four continents. The people I've met are more than scientists: they are explorers. For them, discovering a new element has the same thrill as setting foot in an uncharted land. Already, the superheavy elements are rewriting the laws of atomic structure. They have reached a point where the periodic table loses its meaning; perhaps they will soon end chemistry as we know it.

In the twentieth century, scientists made elements and helped to expand the periodic table. In the twenty-first, they could make the elements that will break it.

This is their story.

Introduction

Kenneth Bainbridge had the worst job in history. As the man in charge of the world's first atomic test, if something went wrong it was his duty to walk up to the bomb and poke it. Bainbridge wasn't a weapons specialist or army officer. He was just a scientist whose research had become very interesting indeed.

In the dawn twilight of 16 July 1945, Bainbridge stood in a cramped wooden shelter deep in the New Mexico desert, the flimsy hideaway protected by concrete and soil. An inhospitable stretch of wilderness used by the US Air Force to train bomber crews, the conquistadors had called this vast expanse of nothing the Jornada del Muerto – the 'journey of the dead man'. It was well named: there wasn't a drop of water that side of the shadowy Oscura Mountains until you hit the Rio Grande.

About 9km (5.6 miles) from the front of Bainbridge's bunker was a pylon. At the top sat the 'Gadget': the first of three nuclear bombs created by the Manhattan Project, the Allied attempt to end the Second World War with a super-weapon. Also in the desert were the project's VIPs: Robert Oppenheimer, head of the scientific effort; Leslie Groves, generally running the show; and James Chadwick, father of the neutron and there to represent the British. Scattered around the desert – most with Groves at base camp 10 miles to the south west or, like Chadwick, watching from a hill some 20 miles away – were other scientists and generals, all anxious to witness the test. The detonation was code-named Trinity.

A countdown began at 5.10 a.m. 'My personal nightmare,' Bainbridge remembered in *Bulletin of the Atomic Scientists* as he recalled the minutes ticking down some 30 years later, 'was knowing that if the bomb didn't go off or hang-fired, I would have to go to the tower first.' It wasn't the first horror he had experienced since joining the project. He had made a similar walk only weeks before, while inspecting a live bomb that had been peppered with high-explosive rounds to see what would happen if the Japanese air defences got a lucky hit. The target had started to smoke, sending Bainbridge sprinting back to his shelter or else risk being blown apart.

There was also the problem of creating a secret camp at the northern tip of a live military bombing range. 'In the middle of May, on two separate nights in one week,' he wrote, 'the Air Force mistook the Trinity base for their illuminated [training] target. One bomb fell on the barracks building which housed the carpentry shop, another hit the stables.' Bainbridge had asked Oppenheimer if, on the next bombing pass, he could return fire.

Then came the lack of understanding from the brass in charge about the true power of a nuclear weapon. Five days earlier, the core of the bomb had arrived at the nearby McDonald ranch house. This was a squat, single-storey adobe home with a few empty rooms – the sort of isolated homestead that gets attacked in western movies. It had been surrendered to the military, and the farmer's boudoir had been turned into a makeshift clean room, a vacuumed lair with the windows sealed by electrical tape. As the core was carried inside, one of the scientists, Robert Bacher, stopped the test in its tracks. Technically, the metallic sphere of pure radioactive death – weighing just over 6kg (13lb), about the size of a softball – was the property of the University of California. Unable to allow several million dollars of the rarest material on Earth to vanish in a nuclear blast, Bacher had turned to the highest-ranking man in the room, Brigadier General Thomas Farrell, and asked for a receipt.

Bainbridge could only watch in horror as the general opened the case and insisted on his money's worth. 'If I was going to sign for it, shouldn't I take it and handle it?' Farrell later recalled. 'So I took this heavy ball in my hand and felt it glowing warm. I got a sense of its hidden power [...] for the first time I began to believe some of the fantastic tales the scientists had told about this "nuclear power".' As he became aware of what he was fondling, much to Bainbridge's relief Farrell stopped playing hot potato with an atomic bomb and signed.

Bainbridge's patience had even been tried the night before the Trinity test. Another of the scientists, Enrico Fermi, had gone around taking bets with the guards on whether the bomb would accidentally set the sky on fire. Bainbridge had been furious – the soldiers didn't know Fermi was (mostly) joking. Yet all those memories paled next to the wait as the test counted down. Finally, a little before 5.29 a.m., the one-minute warning rocket fired. Bainbridge left his bunker and lay down on a rubber sheet, donning welder's goggles and looking away from the blast. The others hit the deck.

Silence.

Then came a blinding flash, a 'foul and awesome display' that faded to ominous purple, green, then white. At ground zero, the explosion carved a hole 1.2m (4ft) deep and 73m (240ft) wide, melting the desert sands into green glassy rock and vaporising any creature nearby. The blast, equivalent to some 20,000t of TNT, was the greatest explosion ever to have been caused by humans.

Most of Trinity's witnesses had been told to lie face down on the ground. Few did. Instead, the observers stood, stunned into silence as they felt the heat of the fireball and bore witness to the world's first mushroom cloud. Half a minute later the blast's shock wave roared past, covering everything in a fine layer of silica dust. As far as 190km (120 miles) away, windows were shattered; for the next few days, army officials had to drive around telling locals that a munitions storage area had accidentally exploded.

It's a popular myth that the first words after the test were Oppenheimer's, quoting Hinduism's holy book the Bhagavad Gita: 'Now I am become death, the destroyer of worlds.' Oppenheimer merely thought of the passage; the person who spoke was Bainbridge.

'Now,' he sighed, 'we are all sons of bitches.'

The Trinity test wasn't just the moment the world entered the age of atomic warfare; it was the moment heavy elements were unveiled to the world. The nuclear core of Gadget, the silvery radioactive sphere Farrell had juggled to the horror of the watching scientists, was forged from a substance nobody else knew existed.

It was called plutonium. And it was made in America.

★★★

The chemical elements are the basic ingredients of the universe. To date, we know of 118 of them, arranged neatly on the periodic table. Our best guess is that there are at least 172 possible elements, even if they have never existed in the universe. If that's true, we haven't found a third of the periodic table's pieces yet. Or, rather, we haven't made them.

All atoms in the universe heavier than lithium – so, all stuff that didn't emerge from the Big Bang – came about through atomic hijinks, either by atoms smashing together, capturing particles or shaking themselves apart. Every night, we can witness the process in action. Stars are effectively cosmic fusion reactors, stellar forges that build all the lighter elements up to iron before, after billions of years, they explode in a supernova that showers the universe with its heaviest components. To create an element is to follow a roadmap that takes us far beyond the reaches of Earth and recreates the processes responsible for our existence.

I'm following a roadmap of a different kind. Today I'm a pilgrim on New Mexico's lonely highways, baking under the sun's heat and seeking a tale of scientific wizardry in the creation of the most destructive weapon in the world. The final 26 elements known to humanity are, for most scientists, irrelevant: most labs don't have a nuclear reactor or costly particle accelerator on hand to create them. The superheavy elements – the final 15, element 104 and beyond – have only ever existed in quantities so small they can't be seen by the human eye and can last for less than a second. They have never been detected outside of the laboratory. They have no known use. A few are so unstable no one has ever managed to conduct an experiment on them. Chances are, unless neutron stars are colliding as you read this, most of these elements do not exist anywhere in the galaxy at present. They are chemistry's unicorns.

I'm on a quest to find out how and why these superheavy elements were made – and what we might be able to do with them in the future. It's the greatest scientific race most people have never heard about. Gadget was its starting gun.

Today the Trinity site is a visitor attraction, a beacon in a scrub desert alive with cactuses, yuccas and lizards. Still part of the US army's White Sands Missile Range, the spot is only open to the public on two days a year. Getting there means a long, dusty drive past giant radio telescopes and lonely settlements. To the west sit a few weather-beaten shacks called Pie Town, and the National Science Foundation's Very Large Array, its dishes pointed at the sky in search of black holes. South lies a town called Truth or Consequences, named after a game show as an April Fool's Day prank, and home to Earth's only spaceport. To the east, Roswell's UFO seekers look for strange things in the sky. In between lies 8,300km^2 (3,200 miles^2) of military isolation.

Breaking from the highway, past a loose band of nuclear protesters, is a straight dirt track. Up ahead, fellow visitors queue to pass the checkpoint into the range, the chrome bodywork of their pickup trucks shimmering in the heat's haze. Here the road whips out into nothing, cutting the same path Bainbridge used the morning he and the other members of the Manhattan Project changed the world. After Trinity, he'd driven back to the barracks in a state of shock, 'swerving off the road only once', where fellow scientist Ernest Lawrence had slipped him a bottle of bourbon. Clutching it to his chest until he could appreciate it, Bainbridge had climbed into bed and collapsed into deep sleep. I don't blame him at all.

The bomb site is another 8km (5 miles) into the desert void. Here, cheery US Navy staff sweltering in digital-blue camouflage fatigues wait to direct the bomb's fan club.

'Aren't you a little far from the sea?' a visitor drawls to the nearest sailor.

'Eh,' she shrugs in reply, too chilled to explain that the range is used by all branches of the armed forces to test America's latest toys. 'There was a flash flood here last week. That's good enough for me.'

Despite the friendliness, Trinity is probably the weirdest tourist trap in the world. Stray off the road and the navy will shoot you. Signs and high wire fences cover the area, warning of radiation, of rattlesnakes, of penalties for pilfering trinitite – the name given to those strange glassy rocks caused by the bomb's fury more than 70 years ago. The mineral, buzzing-hot with radiation, remains scattered across the landscape; the local ants are obsessed with it, and their nests are tracked down regularly with Geiger counters and cleaned out. It's a federal crime to take the rocks away from the site, though plenty of locals sell them by the side of the highway regardless.

There's not too much to see at Trinity. Desert sands filled the bomb crater long ago. The only evidence of a nuclear blast is a small lump of concrete and twisted metal – the remains of Gadget's pylon. Next to it, the military have put up a black obelisk to mark ground zero, its ominous shadow nullified somewhat by the queue to have your photo taken standing at its side.

Guest speakers are on hand to tell the history of the site. One has even brought along a replica of Gadget's 'Fat Man'* design strapped to the back of a trailer. Painted brilliant white – the original was mustard yellow to make debris easier to spot – the replica's swollen belly is as tall as a person, a giant egg with stabiliser fins. It's almost comical, the kind of thing Wile E. Coyote would use to try and finish off Road Runner. Further away sits a hollow pipe of rusted metal, the gaping maw some 3m (10ft) wide. It's all that's left of 'Jumbo', a massive 214t container designed (though never used) to contain a nuclear blast. Now it's yet another spot for an atomic selfie, conveniently located next to a hot dog stand, a barbecue pit and a row of Portaloos.

For a nuclear bomb site, Trinity is surprisingly radioactive-free. An hour's visit is worth about 1 millirem, the radioactive

* Fat Man was the code name for the plutonium bomb's design, which was bulbous because it needed to implode inward to work – something a uranium bomb didn't have to worry about.

equivalent of eating 100 bananas at once.* Its impact, however, will be with us forever. The radioactive material mankind created here has drifted across the world in our atmosphere, raising the background radiation of Earth. Today, all modern steel is contaminated by Trinity and the later atomic tests; the process to make it requires large quantities of air and inevitably sucks in some of this radioactive debris. If steel with low background radiation is needed – such as for highly sensitive Geiger counters – the only option is to use steel made before 1945. Usually, it's plundered from sunken battleships, such as the German fleet that was scuttled after the First World War.

It is only the first example of how elements created in a lab were to change our world.

* All bananas are slightly radioactive, although the lethal 'banana equivalent dose' is eating about 35 million bananas at once. If you did that, you'd have bigger problems.

PART ONE

CHILDREN OF THE ATOM

CHAPTER ONE

Modern Alchemy

There's a temptation to think that creating a new element is alchemy. For hundreds of years, men and women had tried to become a new King Midas. Their dream was to turn lead into gold through 'transmutation', usually through occult rituals and strange experiments. Some were wannabe wizards who bought into the idea of flowing robes and astrological charts. Others were capable scientists who used coded messages to hide their processes as spell books (mystical fluff such as 'take our Fiery Dragon that hides the Magical Steel in its belly' just meant 'use iron'). All of them lived in a world where 'elements' still meant the Greek idea of earth, air, fire and water.

Then, in 1787, Antoine Lavoisier took a major step forward in modern chemistry.* Four elements weren't enough for him. Thanks to his wife, Marie-Anne, who had a knack for translating science books, he knew about the basic materials that weave our universe together. He also knew they couldn't be changed into each other with some magic words. The alchemists were wrong.

Lavoisier set about putting together a list of these 'chemical elements'. A few mistakes crept in (light and calories made his cut) but it was a good start. For the next 82 years, scientists cleaned up, expanded and arranged Lavoisier's ideas. In 1869 a Russian chemist, Dmitri Mendeleev, ordered them into what we call the periodic table.

This was an astonishing feat – the equivalent of putting together a globe-spanning jigsaw puzzle without knowing the picture, the shape or how many pieces came in the box. As

---

* Lavoisier also married a 13-year-old and was executed during the French Revolution for upsetting tobacconists, but such was the life of an aristocrat in eighteenth-century France.

best as anyone could tell, the elements started at hydrogen and worked their way up to the heaviest yet discovered, uranium. Mendeleev ordered his elements by weight and placed them into groups, arranged in columns with similar properties. If the next known element didn't fit his pattern, he left a gap. Gradually, as Mendeleev was proved right and missing elements were discovered, the gaps began to fill. An element's number was arbitrary – just a handy way to keep them in the right sequence.

By the start of the twentieth century scientists knew about the atom. If the elements are the materials of existence, atoms are their tiny building blocks, roughly a billion times smaller than the palm of your hand. Something inside the atom decided which element it made. Thanks to the work of pioneers such as Henri Becquerel and Marie and Pierre Curie, they also knew about radiation. This was the glowing danger, a mystery force of energy emitted from the atom itself. In 1901, at McGill University in Canada, two scientists, Ernest Rutherford and Frederick Soddy, had been playing with the strange phenomenon when they noticed something odd. On the lab bench there had been a small sample of the element thorium. Somehow it had turned into radium.

'My God, Rutherford!' Soddy exclaimed, the Englishman losing his usual cool. 'This is transmutation!'

Rutherford, a pugnacious New Zealander, had hissed him quiet with a withering Kiwi put-down. 'For Christ's sake, Soddy, don't call it transmutation. They'll have our heads off as alchemists.'

This wasn't alchemy – it was the birth of a new science. In 1911, Rutherford discovered the nucleus. This was the tiny core at the centre of an atom; as Ernest Lawrence would later say, if an atom was the size of a cathedral, the nucleus was a fly. Lawrence didn't mention that it would have to be an insanely dense mutant fly that made up 99 per cent of the cathedral's mass.

By 1913 a new model of the atomic world, put forward by Rutherford and the Danish physicist Niels Bohr, began to explain what was happening. Around the nucleus was a series of imagined concentric shells (see Figure 1), each of which could contain a certain number of small, negatively charged particles called electrons.* These interact with other electrons in nearby atoms – the basis for what we now know as chemical bonding – to form molecules. Each element has one more electron than the previous, and is eagerly trying to complete its outer shell.

But the number of electrons doesn't decide the nature of an element. The same year, a young researcher at the University of Oxford, Henry Moseley, added a crucial piece to the puzzle. Using X-rays, he showed that each element had one unit of nuclear charge greater than its predecessor – that something in the nucleus was providing a positive charge that countered the extra electron. Tragically, Moseley never found out how important his work would be; when the First World War

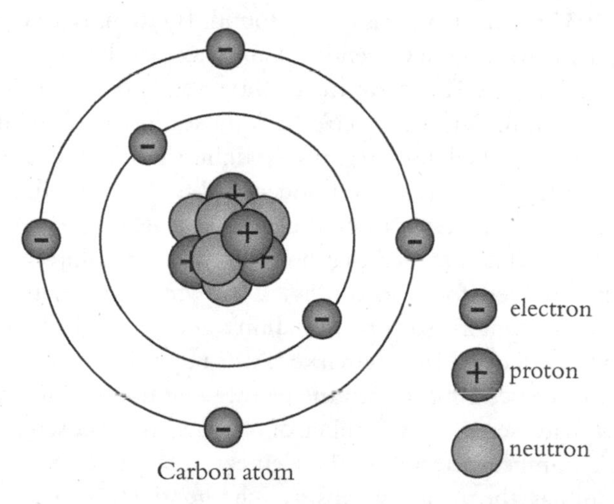

Figure 1 Bohr model of a carbon atom (not to scale), showing a nucleus with six protons and six neutrons, with electrons in two shells.

* Electrons were discovered earlier, in 1897, by Ernest Rutherford's mentor J. J. Thomson. Thomson thought atoms were like plum puddings, with electrons as the raisins.

broke out, he volunteered for combat and was killed at the Battle of Gallipoli. But he had made the breakthrough that explained Mendeleev's elemental order. And five years later, Rutherford had the answer to the mystery of the nucleus.

A nucleus is made up of two main particles. Protons, first found by Rutherford in 1919, decide the element an atom forms. An atom with one proton is hydrogen; two, helium; and so on, up to 92 for uranium. These particles are positively charged, so they constantly try to repel each other: a bit like clenching magnets with the same poles facing each other in your fist. But that still didn't make sense. How come the nucleus didn't just rip itself apart? Rutherford realised there had to be something else inside helping to keep the atom together. He imagined a particle of similar size to a proton, but this time without any charge at all, acting as a kind of packing filler. He called it the neutron.

In 1932 the neutron was finally found. By then, Rutherford had moved to run the Cavendish Laboratory at the University of Cambridge, UK, where his deputy was James Chadwick. Tall and thin, with a raven's-beak nose and stiff manner, Chadwick worked for 10 days straight to find the elusive particle, sleeping only three hours a night. When he finished telling his colleagues all about the discovery, he simply stopped mid-flow and announced that he wanted 'to be chloroformed and put to bed for a fortnight'. Chadwick had twin five-year-old daughters, so probably didn't get his wish. 'It was a strenuous time,' he later remarked, stiff upper lip intact.

With the neutron, the atomic picture and the periodic table jigsaw made sense. The number of protons, also known as the atomic number, determines the element. The more protons an element has, the greater its positive charge, the more electrons it will have (and, with each row of the periodic table, more shells). The number of neutrons in the nucleus creates different versions of the element, called isotopes. If an atom has six protons, it is always carbon. But you can have carbon-12 (six protons, six neutrons), carbon-13 (seven neutrons) or even carbon-22 (16 neutrons). However, the whole thing is a sort of nuclear juggling act. For light elements, the ideal balance is one neutron per proton. As the elements become heavier and their atomic number increases, more neutrons are needed to keep the atom together. But even then, it's usually not enough to make things stable for long. When a nucleus becomes unstable, some of these particles are ejected or can change, which is known as radiation.

Today, we have discovered more than 3,300 different combinations of protons and neutrons (in various energy states) – usually referred to by researchers as nuclides. The majority are radioactive and unstable. Even so, we've barely scratched the surface of what we believe exists. But for the early pioneers such as Rutherford and Chadwick, the idea that so many different combinations of nuclides could exist was unimaginable.

Until, that is, a ragtag group of impoverished Italian scientists decided to shake things up a little.

★ ★ ★

May 1934. Enrico Fermi rushed down the corridor of his lab, racing his student Edoardo Amaldi. Their grey, dirtied lab coats flowed behind them, the floorboards creaking under the strain and the racket of desperate footfalls disturbing the quiet of Fermi's mentor's flat on the floor above. Fermi prided himself on being the fastest physicist in Rome. He needed to be. At the end of the corridor was an experiment where they were trying to create a new element. They had to move fast before their work degraded into something else.

Fermi was the star scientist of Benito Mussolini's Fascist Italy. Short and dark, his face dominated by a cheery grin under a long, thin nose and a large, semicircular forehead, Fermi certainly made an impression. His mind raced, quickly coming up with rough answers to questions that seemed impossible.* When he'd made professor aged just 24, Fermi had set up a lab in the University of Rome's physics department

---

* A classic Fermi problem is 'How many piano tuners are there in Chicago?' By approximating the number of households that had pianos and how often they would need tuning, Fermi was able to come up with a rough estimate of the number of piano tuners the city could support (about 225, for those wondering).

in an old villa off the Via Panisperna in the centre of the city, and had sworn he would drag Italian science kicking and screaming into the twentieth century. He had soon attracted a team of young, fresh talent who didn't mind breaking, bending or ignoring the supposed laws of the universe. None of them, including Fermi, had any real experience in experimental nuclear physics. None of them cared. To them, Fermi was simply 'The Pope', and if he said something could be done, they'd do it. Fermi's 'Via Panisperna Boys' were atomic physics' answer to the Sex Pistols: punk scientists who were about to make their own rules.

Despite Mussolini's patronage, the boys were broke. While their British, French and American rivals could afford the best equipment, Fermi's team had to improvise. Geiger counters were made at home through trial and error. Teenage brothers were bribed to help when something heavy needed lifting. Lacking any protective equipment, the team would hide at the end of the corridor to shield themselves from certain death when dealing with radiation – leading to Fermi's frantic footrace to get his results.

A few decades earlier, Ernest Rutherford had discovered three distinct types of radiation, the first two of which he called alpha and beta. Alpha radiation saw the nucleus emit an alpha particle (two protons, two neutrons – the equivalent of a helium nucleus). This is how Rutherford and Soddy's thorium sample (element 90) had turned into radium (element 88): by losing two protons, it had dropped two places back on the periodic table. The time it took for half of a sample to degrade through radioactivity was known as the half-life.

Fermi was interested in beta radiation. This didn't cause the loss of any protons at all, but turned a neutron into a proton, moving the element one place *further along* the periodic table, and spitting out an electron in the process (the loss of the electron creating the radiation in question).* But the

---

* Strictly speaking, this is beta minus radiation, but that's not really important for our story. Nor is the third main type of radiation, gamma.

equation didn't balance out – not enough energy was being lost. Fermi suggested that something even smaller than the electron was being thrown out of the atom too. Sadly, the journal *Nature* didn't believe him: the paper was rejected for 'speculations too remote from reality to be of interest to the reader'. Infuriated, Fermi decided to do a few silly experiments, just fun in the lab, to take his mind off things. What better way to forget your troubles than to make some stuff glow in the dark? Firing off his thoughts about beta radiation to a few other journals, he decided to move on and do some practical research instead – and a new idea called artificially induced radiation caught his eye.

Bombarding an element with alpha particles resulted in an element *gaining* two protons and two neutrons. In France, Irène Joliot-Curie and her husband Frédéric had turned aluminium (element 13) into phosphorus (element 15) by doing just this; Irène was the daughter of the legendary Marie and Pierre Curie, and radioactivity was the family business. Alpha particles, thanks to their protons, were positively charged, just like the nucleus. This created two problems: first, the negatively charged electrons orbiting around the nucleus would slow them down; second, and worse, they had to be smashed into the target at high speeds to get past the electrostatic repulsion of the nucleus itself – again, just like pushing two similarly charged magnet poles together. To gain entry through this invisible force field, known as the Coulomb barrier, you needed a staggering amount of energy. That required a particle accelerator.

Fermi didn't have one. He couldn't afford one. He had nowhere to put one. Instead, he had a brilliant idea. Why not just bombard your target sample with neutrons instead of alpha particles? They didn't have any charge, so couldn't be repelled; if a neutron hit the nucleus, it was just going to slip into its heart, raise its energy and cause radioactive decay. Perhaps something interesting would happen.

The Via Panisperna Boys had set to work. The first problem was getting hold of neutrons. In the basement of the building, another professor kept a sample of radium in his safe – a single gram, worth almost 20 times the Boys' annual budget. Radium was known to decay (through alpha radiation) into a gas: radon. Unable to steal the sample, Fermi arranged a system of pipes to pump the gas from the safe into small vials, where it was mixed with beryllium powder. Highly radioactive, the radon gas gave off yet more alpha particles; these struck the beryllium atoms, causing them to emit neutrons. Fermi had created a home-made neutron beam. He just needed something to fire it at. Why not fire it at everything?

Fermi wrote out a shopping list covering every element or compound that could be used as a target and gave it to one of his subordinates, Emilio Segrè (Fermi, according to his wife Laura, 'disliked shopping intensely'). He then skipped off home for lunch and a nap, while Segrè set off around Rome trying to track down 'The Pope's' extensive demands.

Segrè was a cornerstone of Fermi's schemes. The son of a paper magnate, he had initially trained as an engineer and spent two years in the Italian army working on anti-aircraft guns before switching to physics. Following Fermi's instructions, he soon cornered a supplier, but the man only spoke a regional dialect and, having been raised by priests, Latin; fortunately, Segrè knew enough of the dead language to make himself understood. The chemical shopping list was soon filled; the supplier even gave away some unused stock, high on dusty shelves, for free. Nobody had asked for those elements in 15 years.

The team tried their targets in order of their atomic number. The first few elements didn't really do anything. The next gave off alpha particles, moving them two places back on the table. But the heavier elements didn't give off alpha particles so easily. Instead, the force that kept the nucleus bound together – the nuclear strong force – held onto the neutron. This caused the element to eventually undergo beta decay, creating an element one place further along the periodic table. Fermi had stumbled on a process called neutron capture.

Back to Fermi's frantic sprint down the corridor. After weeks of study, the team had finally reached the end of the periodic table. They were about to see what happened to the heaviest element: number 92, uranium. Rounding the door ahead of Amaldi, Fermi slipped in first to grab the samples and check the results. Quickly, the Boys worked to analyse what they had found. The uranium had transmuted into something else. The only question was which way it had gone: alpha decay into a known element, or beta decay into something new?

The team compared the new substance's half-life and chemistry against the known elements as far back as lead (element 82).* If alpha decay had occurred, it would have started a 'decay chain' – turning into element 90, then 88, then 86 and so on. But there was no sign their creation had undergone alpha decay at all. The only explanation the team could think of was that the sample had undergone beta decay. Further tests by the Boys soon found a second mystery element. Again, the only explanation was another beta decay.

Up until that moment, most physicists didn't believe there could be an element heavier than uranium. It was the end of the periodic table. It had been known about since its discovery in 1789, while Lavoisier was still alive and before the periodic table even *existed*. Fermi and his chaotic squad of rebels, using home-made gadgets and dusty targets given away for free, had done something the greatest laboratories in the world, using the greatest machines of the day, hadn't even thought possible. They had just broken the boundaries of Dmitri Mendeleev's table of elements.

Fermi told the world he had created elements 93 and 94.

★ ★ ★

It's 2018, some 78 years after Fermi's announcement, and I'm stood in the heart of Rome. If I'm going to understand why

---

* Element 85, astatine, hadn't been discovered at the time, but Fermi knew it couldn't have been produced by uranium, as alpha decay loses two places at a time, so would have skipped over it.

the elements matter, I need to step back to the first attempts to push beyond uranium. And that means trying to find a trace of Enrico Fermi.

The Eternal City doesn't change. Rome is still as Fermi would have known it, a hilly maze of effortless disorder somehow manicured into one of the world's greatest capitals. It is a hive of accented shouts from restaurants, apartments and taxis; of crumbling buildings coated in graffiti and licked with flyers; of tight alleyways that suddenly twist you into sweeping, open avenues lined with marble columns built to honour long-dead emperors. No other city can pull off disrepair quite like Rome. No other city can exude such lazy, sun-drenched style. No other city could have produced Fermi.

It's a long journey from New Mexico, and a strange transfer from barren wilderness to the kinetic pleasures of a European capital. The Via Panisperna still exists, winding down from the towering Basilica of Santa Maria Maggiore, its route cutting a bumpy tumble over the Esquiline and Viminal hills toward the city's ancient heart. Part of the trendy, artisan quarter, the road's cobbled surface has been beaten flat by Fiats and Vespas, and its buildings house fashionable hair salons, quiet bistros and high-end art galleries. A short walk away, visible from the end of the street, the Colosseum rises up in the distance. I stop and catch my breath, admiring the colonnades of the ancient gladiatorial fighting pit as they are rendered by the setting Latin sun.

This is as far as I can go. Fermi's former lab is behind a wall 15m (50ft) high, tucked into what is now the Italian Ministry of the Interior. The entrances to the complex are guarded by the Carabinieri police force, and the façades are covered in CCTVs to discourage anyone with a bright idea. All I can do is sit on the steps leading up to the villa, eat an ice cream and marvel at how Rome's unique attitude to the past means its greatest scientific son doesn't even get a plaque. Perhaps that's what happens when you have 2,500 years of history to spare.

Back in 1934, the Via Panisperna discovery of 'transuranic' or 'transuranium' elements (elements beyond uranium) became an overnight sensation. There were a few doubters – most notably the German chemist and element discoverer Ida Noddack, who thought that the nucleus had split apart into much smaller elements – but nobody listened. Mussolini's press proclaimed a triumph as great as ancient Rome, boasting of 'Fascist victories in the field of culture'. 'One second-rate paper went so far as to state that Fermi had presented a small bottle of [element] 93 to the Queen of Italy,' Laura Fermi recalled in her biography of her husband, *Atoms in the Family* (he hadn't; Fermi knew all too well that Marie Curie was dying of cancer caused by radiation, and that carrying a vial of radium in her breast pocket to show off at parties probably hadn't been her greatest idea). A proud Mussolini pressured Fermi to name his elements after the lictors of ancient Rome, whose symbol, the *fasces*, gave fascism its name; uncomfortable with the attention, Fermi instead chose 'ausonium' and 'hesperium', after the Greek names for Italy.

While he worked to confirm his findings, Fermi stumbled onto something even more exciting. If they put something in the way of the beam that reduced the speed of neutrons, such as paraffin wax, they got much better results. Fermi realised it was the hydrogen atoms in the molecules of wax slowing down the neutrons, and reportedly yelling 'Fantastic! Incredible! Black magic!', he snatched up the equipment, ran downstairs and leaped into the institute's fish pond – bathing the device in the most hydrogen-rich environment he could find: water. Fermi had just uncovered the principle that powers today's nuclear reactors. Slow neutrons mean they spend longer near the nucleus, and thus are more likely to get captured – resulting in more reactions. Water was the perfect way to slow them down. When the friends went to type up their discovery later that evening, the whole lab was in such a good mood the maid assumed they were drunk.

Fermi's idea of using slow neutrons to attack the nucleus was soon copied by bigger teams with better machines to see if they could make new elements and isotopes too. The Americans, French and Germans threw themselves into experiments in a race to fill in all the missing spaces in the chemical world: it was new, exciting science that offered a host of possibilities. In 1937 Segrè, who had struck out on his own in Sicily, asked for spare parts from one of the American 'atom smashers' at the University of California, Berkeley. Investigating a filter sent from the Berkeley team, he found they had created the missing element 43 without noticing. Segrè and his collaborator were credited with the discovery. Italian punk science was filling up the periodic table.

Yet the smiles and happiness soon vanished. The days of the Via Panisperna were over. As Europe tumbled toward war, Adolf Hitler's Nazi Germany fuelled a climate of anti-Semitism that Mussolini was only too eager to embrace. Fermi and the Boys realised just who their patron was – and knew they had to get out of Italy.

In 1938, while Segrè was on a trip to the US to further explore his element 43, Mussolini decreed that 'Jews do not belong to the Italian race' and barred any member of the religion from holding a university position. Segrè was a Sephardic Jew and had married a Jewish refugee from Nazi Germany. Wisely, the couple decided to cut their losses and make California their new home.

Laura Fermi was also Jewish. If Enrico Fermi's wife and children were to be safe, he knew they would all have to flee Italy, taking only what they could carry. Already, he had spied a possible escape route. Earlier that year, while attending a conference, he had been taken aside by Niels Bohr and told he was being considered for the Nobel Prize. With it, Fermi knew, would come enough prize money to start a new life in the US. All the Italian could do was wait and hope the Royal Swedish Academy of Sciences, who judged the award, agreed with Bohr's assessment.

On 10 November 1938 the Fermis woke to the news of *Kristallnacht*. In Germany, Jewish families had been beaten and killed, their shops smashed and their synagogues burned to ash. Shortly after, the telephone rang: Fermi was told to expect a phone call from Stockholm, home of the Nobel Committee, at 6 p.m. Fermi, making one of his rough calculations, estimated his chance of receiving the world's greatest prize at 90 per cent. He took his wife to buy a pair of expensive watches. Knowing his hatred for retail therapy, Laura understood he was turning their money into something a refugee could pawn. Together, hand in hand, they walked the streets with no real destination in mind as the hours ticked down. It was a silent goodbye to their old life.

That evening, Laura listened to the 6 p.m. news on the radio. Anti-Semitic laws were being introduced across Italy, limiting rights and freedoms, excluding Jewish children from schools, confiscating passports. Her heart sank. Then the phone rang. Enrico Fermi had won the Nobel Prize, on his own, 'for his demonstrations of the existence of new radioactive elements produced by neutron irradiation, and for his related discovery of nuclear reactions brought about by slow neutrons'.*

It was deliverance. Without a moment's hesitation, Enrico accepted a trip to Sweden to collect the medal in person, then humbly agreed to a 'temporary' lecture position in the US. Naturally, his family accompanied him. By the time the Fascists realised what was happening, the Fermis had slipped out of Europe and were on the next liner to New York.

It was there Fermi learned of shocking news: he and the Boys had been completely wrong. Worse, they had missed out on the discovery of the century.

★ ★ ★

The Nobel Committee has made some howling errors in its time. In 1949 it gave a prize to the man who invented lobotomies, a practice now banned across the world. In 1926 it awarded a medal to Johannes Fibiger, who thought parasitic

* Fermi could also have won for the paper on beta radiation *Nature* had rejected: it described the possibility of a 'baby neutron' – or, in Italian, *neutrino*. This was eventually found in the 1950s, and gets name-checked in pretty much every episode of *Star Trek*.

worms caused cancer (they don't). And in 1938 it honoured Enrico Fermi for creating the first elements beyond uranium. He hadn't.

One of the great problems when it comes to making new elements is how to prove you've actually done it. It's perhaps the defining problem of element discovery: you're dealing with something nobody has ever seen before. Fermi had made an honest mistake. If he had checked his results a little more thoroughly, he'd have seen that the 'new elements' were just barium, krypton and other atomic debris.

If he had realised at the time, Fermi would have been far from disappointed – he probably would have done cartwheels in delight. Never mind new elements, the Via Panisperna Boys had done something even more miraculous: they had made an atom *explode*.

The discovery ended up falling to a German group. Otto Hahn was working at the Kaiser Wilhelm Institute for Chemistry in Berlin. In late 1938 Hahn and his assistant Fritz Strassmann were experimenting with slow neutrons and uranium, much as Fermi had done. However, the results were confusing. No matter how hard he tried, Hahn couldn't make element 93; instead, he was producing barium, element 56, best known today for its use in enemas. The only way it could have been made was if he had somehow split a uranium atom in two.

Unlike many of his colleagues, Hahn was not a Nazi. A few months earlier he had helped his old lab partner, Lise Meitner, escape the country, going so far as to give her his mother's diamond ring in case she needed to bribe the border guards. Convinced he was going wrong somewhere and stuck for an answer, Hahn wrote to Meitner: 'Perhaps you can suggest some fantastic explanation ... we understand that it really can't break up into barium.'

Meitner was in Kungälv, Sweden with her nephew, the physicist Otto Frisch. Both Jewish, the duo had been displaced by the Nazi regime (Frisch's father was imprisoned in a concentration camp), and friends had invited the lonely scientists to stay for Christmas. Figuring that atomic physics was more fun than charades, they decided to work on Hahn's problem together.

The pair of exiles soon had an answer. By the late 1930s the nucleus wasn't viewed like a brittle core that could be cracked and split – it was imagined as a high-density drop of liquid. Adding a neutron made the liquid shake. Meitner took a pencil and started drawing. What if the energy made the liquid turn into a dumb-bell shape? At some point, the forces binding the nucleus together and the electrostatic repulsion trying to pull it apart would cancel each other out. The droplet would goop, split and create smaller droplets.

Ida Noddack, the chemist who had doubted Fermi's discoveries, had been right all along. Creating an element was walking a tightrope: you needed energy to push past the Coulomb barrier in the first place, but if you used too much energy the nucleus would literally shake itself apart. In other words, the nucleus acted like a fuse box: too much energy and the fuse blew. Meitner wrote back to Hahn. They called the discovery 'nuclear fission'.

It was the biggest discovery since the neutron: something that Fermi, the British, the Americans, everyone, had missed completely. Overnight, Fermi's elements 93 and 94 vanished as the scientific world woke to the consequences of Meitner and Hahn's discovery. By breaking apart the nucleus of the atom, some of the massive energy that bound it together was released. A single atom undergoing fission wouldn't do much. But if you could get a lot of atoms to fission at once, one after each other in an irresistible chain reaction, the power would be immense. If you could control it, the energy could light a city … or flatten one.

Fermi's experiments had spurred a competition to create and understand new elements. Meitner and Hahn's discovery meant nuclear science was no longer a matter of academic curiosity. It was a race to make the first atomic bomb.

CHAPTER TWO

The Secret of Gilman Hall

Luis Alvarez needed a haircut. Taking a mid-morning walk across the hilly rise of the Berkeley campus, the 27-year-old physicist, blond locks tousled in the breeze, admired the view out across the San Francisco Bay. In the distance he could see the new federal prison on Alcatraz; past that, the recently constructed suspension bridge across the Golden Gate, the longest and tallest bridge in the world. Reaching the barber, he settled into the chair and pulled up the morning's *San Francisco Chronicle* with immense satisfaction. It was Tuesday, 31 January 1939.

Alvarez stared at the page in shock. The *Chronicle* had picked up Hahn and Meitner's discovery of nuclear fission from a wire service. Stopping the barber mid-cut, Alvarez snatched the page, leaped from the chair and sprinted back to the Radiation Laboratory. This was at the heart of the campus, a small, ugly wooden hut sandwiched between the grand, Beaux-Arts colonnades of LeConte Hall and the rising needle of the campus clock tower, modelled after the Campanile of San Marco in Venice. Inside, the Rad Lab was a chaotic mess, full of cages of mice for biomedical experiments, blackboards chalked with equations and giant magnets for the 'proton merry-go-round' that was the Berkeley particle accelerator. It was probably the most advanced laboratory in the world.

Skidding through the door, Alvarez hurried to his assistant, Phil Abelson, to tell him the news. 'I have something terribly important to tell you,' Alvarez relayed. 'I think you should lie down on the table.' Abelson was game and, grinning, climbed up on the workbench, spreading himself out among the chemical reagents and machine tools. *Fission! Atoms can split apart!* Abelson went numb; he'd been experimenting with uranium, had noticed similar results to

Hahn's and was probably moments away from making the discovery himself.

Alvarez didn't stop, rushing around to tell everyone, not caring that he had accidentally invented the mullet. He started herding anyone nearby, including future Manhattan Project leader Robert Oppenheimer, to show them the fission energy 'kicks' coming off the lab's own machines. For years, the 'atom smashers' had been picking up strange radioactive noise, but it was always dismissed as a quirk of the machine. The Berkeley team had all missed the discovery of the century – and it had been staring them right in the face.

By nightfall, news of the breakthrough had reached the journal club, where a 26-year-old researcher called Glenn Seaborg heard the tale. 'I walked the streets of Berkeley for hours,' Seaborg later recalled in his autobiography, *Adventures in the Atomic Age*. 'My mood alternated between exhilaration at the exciting discovery and consternation that I'd been studying the field for years and had completely overlooked the possibility.'

Seaborg was a chemist who had been adopted by the physics set. Immensely tall and lean, with a wide grin and pocked cheeks, the young man hailed from Ishpeming, an icy backwater in the wilds of north Michigan, not far from the Canadian border. His family as far back as his great-grandfather were machinists from Sweden and were hardy stock used to rolling up their sleeves. When his grandfather had passed through Ellis Island in 1867, an immigration official had Anglicised 'Sjöberg' to 'Seaborg' with a wave of his pen. There was no way the official could realise that his invented surname would, 130 years later, be inked indelibly into human history.

Seaborg had grown up speaking Swedish at home, scraping out a life in a town whose 'unpaved streets were tinged red from iron ore'. It was a hard upbringing; Ishpeming's only claim to fame was hosting the first away game for the newly formed Green Bay Packers American football team (the Michigan men had sent three of the Packers off with broken bones in the first three plays). When he was 11, the family had moved to Los Angeles, where the young Seaborg's world had suddenly expanded into technicolour. In Ishpeming he'd never heard a radio or seen a building more than a few storeys tall; in the City of Angels he'd found his world filled with lights, cars and oil riches. Inspired by the glitz and glamour of Hollywoodland at his doorstep, the young Glen had added an extra *n* to his name – it just looked cool – and had gone to find his fortune in science.

Science isn't made for fortune hunters. Seaborg had struggled, paying his way through the University of California, Los Angeles first by working at factories and as a lab assistant, then by borrowing money from high school friends. Eventually, he had ended up at Berkeley. In 1937, in one of those quirky moments of luck that change history, he'd been strolling about campus when one of the Radiation Laboratory team had asked him to help with the tricky task of separating out different elements in a solution – he was literally the first chemist they could find. After that, he had been as much a part of the lab as any of them, covering the chemistry while the physicists created new radioactive isotopes of tin, cobalt, iron and Emilio Segrè's element 43 (later called technetium). Many of these isotopes, such as technetium-99m and cobalt-60, were to become the cornerstone of modern radioactive medicine; today they are still used in millions of cancer treatments and diagnostic tests around the world.

The secret to the Berkeley Radiation Laboratory's success was its revolutionary approach. The opposite of the Via Panisperna Boys' homespun charm, the Americans were all about 'Big Science': big teams using big equipment from sponsors with big pockets. It was the brainchild of Ernest Lawrence, the grandson of Norwegian immigrants from South Dakota, who had broken through the stuffy, elitist physics department at Yale before moving west to make his name. Under his guiding hand as director, the Rad Lab had become a template for modern research. Rather than each scientist working on their own experiments, laboriously blowing their

Figure 2 The Berkeley 60-inch cyclotron, a particle accelerator designed by Ernest Lawrence.

own glass tubes, making their own circuits and testing their own reactions, Lawrence expected everyone to work together. Berkeley was an epicentre of large-scale partnerships, teams working in shifts and each member focusing on their own area of expertise.

Throughout the 1930s, Lawrence had pioneered the construction of compact particle accelerators called cyclotrons, the most powerful research machines in the world. It was the spare parts from one of these behemoths that Segrè had cheekily borrowed to discover technetium. Unbeknown to their operators, the machines had already produced another element too.

Its discovery came hot on the heels of Alvarez's barbershop dash. Edwin McMillan was a California native who had made it as far as Princeton before being lured back to his home state by Lawrence. Still in his early thirties, Ed was one of the cyclotron's best operators and possessed a keen experimentalist mind that wouldn't rest until it had answers.

With fission, the material doesn't stay in one place: like a miniature nuclear bomb, the atom blows itself apart, scattering itself in all directions.* Following the revelation that fission occurred, McMillan decided to run some experiments to see just how far things pinged away. Heading to the nearest atom smasher, he started pummelling a sample of uranium trioxide. Soon, he got a very strange reading. Something had been left near his original target material. It wasn't uranium, and it hadn't flown off far enough to be a fission product. Weirder still, the unknown radioactive lump had a half-life of 2.3 days, which didn't match anything previously recorded. Was it the real element 93?

McMillan was flummoxed, so he asked Segrè – comfortably settled in California – to investigate. The Italian was a bad choice for a lab partner. Following the columns down the periodic table, element 93 was supposed to behave like the elements in group 7. Instead, it behaved like a group called the rare earth metals, or lanthanides – a line of elements beginning at lanthanum that all acted similarly to each other. These elements were so odd they had been considered apart from the rest of the periodic table, isolated on a naughty step just below the main chart (to this day, they are almost always displayed apart from the main table). Failing to learn from the slapdash approach to chemical confirmation during his days under Fermi, the Italian decided it wasn't anything important and told McMillan to forget about it. He even went as far as to publish a paper: 'An Unsuccessful Search for Transuranic Elements'.

McMillan wasn't so sure. If his discovery was a fission product, why hadn't it been scattered off like everything else? If it was an unknown isotope of uranium, why didn't it behave like one? The result gnawed at the back of his mind. Over the winter, he tested his mystery sample with hydrofluoric acid and a reducing agent. The result – using chemistry so simple a student could do it – ruled out any of

---

* Things don't stay in one place with alpha or beta decay, either; part of the big challenge in nuclear physics is nothing ends up where you want or expect it to go.

the rare earths. As he worked, life continued. The Second World War broke out in Europe; *Gone with the Wind* and *The Wizard of Oz* were released; the campus swung to the beats of Glenn Miller and Billie Holiday. In May 1940 Abelson returned to Berkeley on holiday (he'd since graduated and moved to Washington, DC), and McMillan asked for a second opinion. In two days, Abelson had performed the full chemical work-up Segrè hadn't.

The results were conclusive. Edwin McMillan and Phil Abelson had discovered the real element 93. The duo published their findings in the *Physical Review.* The Second World War meant there was no fanfare; the great minds of Britain, France and Germany were at war. The only research into fission seemed to be coming out of Russia, where a pair of young physicists had proved that elements can spontaneously fission in nature. Instead of the physics world eagerly discussing their discovery, they received an official protest from James Chadwick and the British, currently under siege as they prepared for the Battle of Britain: *Would you mind awfully just shutting up about things the Nazis might find useful?* Abelson headed back east; McMillan kept working.

The person following McMillan's progress the closest was Seaborg. Both men lived only a few rooms apart, and the chemist pursued the element maker around campus asking questions: in the lunch hall, in the corridor, even into the shower room. Seaborg was hooked, in love with the idea of a new element, desperate to know every detail. McMillan happily took Seaborg into his confidence about his latest scientific escapade. Using a cyclotron, he was bombarding uranium with deuterons (an isotope of hydrogen, with one proton and one neutron) to try and make an even shorter-lived isotope of element 93. His hope was that this more unstable version would undergo beta decay, turning into element 94. Things seemed to be going well. Then one day, McMillan was gone.

Seaborg soon found out why. The US was preparing to join the war, and Lawrence had been asked to give up his best scientists to the military. Along with Alvarez, McMillan had been sent to Boston to work on developing radar detection. Not willing to surrender his new passion, Seaborg wrote to McMillan asking if he could take over the project. Seaborg later recalled in his autobiography: 'Ed wrote back immediately to say he had no idea when he would return to Berkeley and expected a long absence, so he would be happy for us to continue.'

The chemist wasn't going to pass up his opportunity.

★ ★ ★

A neutron walks into a bar and orders a drink. 'How much?' the particle asks. The barman shakes his head. 'For you, no charge.'

The oldest joke in atomic physics is proudly emblazoned on the menu at the Berkeley chemistry department's coffee stand. I'm sat just outside, fighting off the jetlag and wind chill with caffeine and carbs. I thought California was supposed to be sunny? There's a nip in the air this morning and I'm beginning to regret not bringing a thick sweater; part of me wants to give in, dart into one of the town's countless clothing and thrift stores and get a hooded top – probably a goofy one with 'I heart San Francisco' all over it. For now, hot coffee will do.

Berkeley is a small town that morphs seamlessly into bustling Oakland, a relaxed suburb filled with counterculture shops, cheap eats and bars proudly pledging allegiance to its Golden Bears sports programme. The University of California campus dominates the whole area, its manicured lawns and imposing halls resting on a rise that slowly steepens, building to a climax at Grizzly Peak. Originally named after an Irish bishop who didn't believe in the material world, Berkeley has always been home to the raucous and the radical – Ginsberg to Green Day – its left leanings evident on every lamppost or window plastered with slogans such as 'Occupy' and 'Resist'. In the 1960s the Bay Area was the epicentre of the 'flower power' movement and anti-Vietnam protests; today it wears its campaigns for LGBT rights and an end to pseudoscience with righteous pride.

This is one of the world's great science hubs. Starting with Lawrence's Rad Lab, Berkeley has been on a roll of Nobel Prizes and ground-breaking science for 80 years. Just strolling through campus, you could bump into George Smoot, one of the world's leading experts on the Big Bang, or Jennifer Doudna, the biochemist whose CRISPR discovery could allow us to rewrite our genes. The whole complex has a collegiate air, a sense that something very smart is going to happen, mixed with breezy cool and a hint of mischief. Quite what Bishop Berkeley would have made of it is anyone's guess.

LeConte Hall and the Campanile still stand, though the old Rad Lab has since been torn down. Apparently, someone finally realised that conducting radioactive experiments in the centre of one of the world's busiest campuses was a bad idea. Today its descendant, Berkeley Lab (officially Lawrence Berkeley National Laboratory), sits atop the sharp rise behind the university grounds, accessible either by a stiff hike or a convenient shuttle bus.

I'm not here for the lab today; that's a mission for another time. I'm here to break into Gilman Hall. Gilman is another of the beautiful, grey-stone buildings on the campus where Lawrence's men cut their teeth. In 1940, Seaborg had recruited two collaborators, Joseph Kennedy and Arthur Wahl, to continue McMillan's hunt. Aware that their discoveries could be used for an atomic bomb, the team had sworn to work in secrecy. To complete the chemical separations required to explore new elements, the team needed space away from prying eyes. The third floor of Gilman was the best they could come up with.

Sneaking in through the large oak doors that guard the hall's main entrance, it's easy enough to slip onto the staircase – wonderful, heavy concrete stairs with sturdy metal banisters and wood lacquered to a sheen – and climb up to the top floor. Here are attic rooms, still in use by the chemistry department as offices. In Seaborg's day they served as miniature lab spaces with industrial sinks and workbenches, with bottles of reagents, hand-blown glass beakers and retorts fighting for space against Bunsen burners, jars of powders and char-blacked draining boards. Cramped and cosy (particularly for the tall Seaborg), it's the opposite of the vast underground lair Hollywood has associated with scientific breakthroughs. Down a whitewashed corridor, exposed pipes humming overhead, next to an emergency shower for chemical mishaps, you'll find a hardy door. On the wall are two plaques that hint at what came to pass in the locked chamber. Room 307. The place where Glenn Seaborg isolated plutonium.

In mid-December, following McMillan's plan, the trio created the new isotope of element 93. It was a vigorous beta radiation emitter, meaning it was likely turning into yet another element. But they had also made something that fizzed with alpha particles instead. Could 93 have beta decayed and given birth to an atomic daughter?

Work continued through a long, wet winter. Room 307 soon stank with reagents and reactions, forcing the team to open the windows and work out on the balcony. Here, in full view of the world, three men in their twenties played with perhaps the most secret substance ever to have existed. Much of the work was carried out at night, the conspirators making desperate runs with their radioactive samples between the Rad Lab machines and the Gilman lair. All that was missing was the final confirmation that element 94 was real.

The breakthrough came as a cliché. Every scientist knows that, sometimes, the best results come when everyone else has left and you're trapped alone in the lab. On the night of 21 February 1941,* a wild storm battered and bruised the San Francisco Bay. Wahl was still in the attic space, the whole room rattling with the wind and rain, lightning flashes occasionally illuminating the downpour outside. A little past

* The plaque outside 307 insists that the discovery was the night of 23 February. Seaborg always claimed it was during the storm, and that's good enough for me.

midnight, his eyes growing heavy, Wahl finished his last chemical test. Chemists are obsessed with oxidation numbers – how many electrons lost or gained by an atom when it forms a compound. The group had just proved that the radioactive daughter particle had a higher oxidation number than any known element it could have been. It *had* to be element 94.

Standing in the hall, the place resonating with history, I can't help but imagine Wahl laughing maniacally as the balcony doors burst open and the thunder pealed behind him. It's probably the only time in scientific history mad scientist chic would have been appropriate.

★ ★ ★

To make a nuclear bomb, you need a chain reaction – one atom going off won't release enough energy to make a big enough bang. This requires a 'fissile' isotope – one that, if hit by a neutron, will send neutrons flying out in turn, like a pool ball hitting the stack. These neutrons will hit other atoms, causing them to explode, which will send more neutrons out, causing yet more explosions, sending yet more neutrons out, etc. If you have enough fissile material – the critical mass – you get a sustained nuclear chain reaction. One atom undergoing fission could flick a speck of dust; 6kg (13lb) of atoms undergoing fission almost simultaneously could level a city.

As far back as late 1939, at the request of Albert Einstein,* President Roosevelt had already put together an advisory committee to consider the feasibility of an atomic weapon. There was one obvious choice of fissile material to make the bomb: the naturally occurring uranium-235. Most natural uranium is U-238, but this was easy to mine, and could then

* The letter was actually written by Leo Szilard, but was signed and sent by Einstein – and if *Albert Einstein* tells you something is a good idea, you should probably listen.

be enriched by a process called gaseous diffusion to remove some of the unwanted isotopes and increase the concentration of U-235.

The second possibility was Seaborg's suspected element 94. By the summer of 1941, Seaborg's team had completed the final hurdles surrounding their creations. Even before Wahl's final tests, the group became aware that at least one isotope of their creation might be fissile. Here, a familiar face joined the party. After dismissing McMillan's discovery, Emilio Segrè had been working with a different group and discovered yet another element, the missing 85, later named astatine. Lawrence asked Segrè to partner up with Seaborg to see if element 94 could make a bomb.

Seaborg wasn't happy. Segrè was a lousy chemist, and as an Italian he was classed as an 'enemy alien' and was not allowed to know the details of what was going on. The situation was crazy; Seaborg would gather chemicals and tell Segrè what to do, but could not tell him what substances he was dealing with or why he was using them. But Segrè had contacts Seaborg could only dream about. When the team needed a larger amount of uranium to bombard, Segrè made a call to Enrico Fermi, who had settled on the East Coast. Soon, 5kg (11lb) of uranium arrived at Berkeley with his former mentor's compliments. With it, the unlikely duo worked out that you'd need to bombard 1.2kg (2.6lb) of uranium in a cyclotron to get 1 microgram (or μg – a millionth of a gram) of element 93, which degraded quickly to 94.

Now with enough element 94 to play with, Seaborg and Segrè soon determined that one of its isotopes was fissile. The team estimated that its fission rate was 1.7 times that of uranium. Element 94 was not just *an* option for a nuclear bomb. It was *the best* option.

Lawrence was a member of the advisory committee and sent Seaborg across the US to explain what he had found to Arthur Compton, the committee member tasked with writing the report on whether a bomb was feasible. Compton listened, but decided against telling the president about Seaborg's new element. On 6 December 1941 the advisory committee met and decided they would proceed with making a nuclear bomb using uranium-235.

After the meeting Compton went to lunch with two of the committee members and brought up the topic of element 94 as an alternative to uranium. He had been talking with Lawrence and other scientists who had convinced him that the Berkeley discovery was worth investigating further. 'Seaborg tells me,' he informed his companions, 'that within six months from the time [94] is formed, he can have it available for use in the bomb.'

One of the diners was James Conant, the president of Harvard University. The New Englander practically sneered at the suggestion. 'Glenn Seaborg is a very competent young chemist,' Conant remarked, 'but he isn't *that* good.' Still, the group agreed it was useful to have an alternative option for a bomb just in case the US found itself at war.

Twenty-four hours later the Japanese attacked Pearl Harbor.

CHAPTER THREE

How to Build a Nuclear Weapon

In the summer of 1941, while Seaborg was flying across the country trying to sell his new element, he had someone on his mind. For the past year he had regularly bumbled into Ernest Lawrence's office, always on some flimsy pretext, just to talk to his secretary Helen Griggs. Griggs, still only 24, had captured Seaborg's heart. An orphan born in a home for unwed mothers in Sioux City, Iowa, her adopted parents had given her a good upbringing and outstanding work ethic. When her father died, she and her mother had moved out west to California, where she'd worked several jobs to put herself first through an associate degree at a local community college, then university. While studying for her degree she'd started working in the Rad Lab's secretarial pool; when she graduated in 1939, the role became full-time. One of her first tasks was trying to persuade her boss to accept the Nobel Prize: Lawrence was in the middle of a tennis match at the time and, unlike Enrico Fermi the year before, had the luxury to decide that the call from Stockholm could wait.

As Lawrence's secretary, Griggs already knew about the secrets of the Gilman Hall attic – she was the one typing up the team's reports. She also knew Seaborg had broken up with his previous girlfriend because he was playing with new elements when he was supposed to be out playing with her. ('She'd taken that as a sign of my priorities,' Seaborg sheepishly recalled in his autobiography, 'which I guess it was.')

Griggs had a soft spot for the tall, awkward chemist. When Seaborg's friend Melvin Calvin told her he was picking up Seaborg from the airport, she agreed to tag along in his Oldsmobile convertible. Calvin was aware of Seaborg's infatuation; he was fed up with his friend's pining and decided to play Cupid. Griggs was more than happy to play along.

The Oldsmobile wasn't a subtle car; its bonnet was almost as long as the cab itself, with smooth art deco lines that oozed luxury. Landing at Oakland from the sleeper flight, Seaborg spotted Griggs in the passenger seat and his heart leaped. Seaborg took the wheel – he paid Calvin's insurance for the right to use the car whenever he liked – dropped his friend off at Berkeley and took Griggs for a drive through the golden hills of California's wine country. Past the peaks that surround the San Francisco Bay, the world opened up to a hot, lazy land of small farming communities as far as Livermore. As they drove, Seaborg talked: finally, he had someone he could open up to about his secret work. His new girlfriend listened enraptured. The couple were made for each other.

When the US entered the Second World War after Pearl Harbor, Griggs was there to support her boyfriend. Arthur Compton took responsibility for the element 94 part of the bomb project, and Seaborg began to feel pressure to deliver the material he had promised. 'Before, we'd been jogging toward our goal,' Seaborg recalled, 'now we hit a dead run.' In early 1942 Seaborg received his orders. Compton wanted to bring all of his scientists together at the University of Chicago's newly established Metallurgical Laboratory. Seaborg was moving east.

The evening Seaborg learned about his move to Chicago, he took Griggs for fried chicken at Tiny's Waffle Shop to ask her to come with him. Forever bashful around her, he promptly lost his nerve. The couple headed back to her apartment, where Seaborg again tried – and failed – to express himself. Finally, he just came out with it: 'I'm sitting here trying to think of a way to ask you to marry me.'

Griggs said yes.

The news shocked Ernest Lawrence (the romance had been a bigger secret than element 94), but he agreed to let her go. Soon, rumours circulated around the Rad Lab that Seaborg had only proposed because he needed a good secretary. The couple didn't care. In April 1942, on his thirtieth birthday, Seaborg arrived in Chicago. Two months later he returned for Griggs. They left California, stopped at the first town across the border in Nevada to get a 'quickie' marriage, and then headed to their new home.

The two new elements had been named only a few months earlier, in March 1942. At first, the team had used code words to throw anyone off the scent. Element 93 was 'silver'; 94 was 'copper'. That had worked until the group needed to talk about *actual* copper, which became 'honest-to-God copper' instead. But as the discoveries became more widely known, continuing to call them 'silver' and 'copper' became too confusing for everyone involved. They needed to come up with alternatives.*

Seaborg and McMillan drew up a list of options. They considered 'extremium' or 'ultimium' – after all, their finds had to be the end of the periodic table, didn't they? – but eventually took inspiration from the solar system. When uranium had been discovered in 1789 it had been named after the planet Uranus; McMillan decided to copy the idea and call element 93 'neptunium', after Neptune, the next planet in sequence. Seaborg followed suit and element 94 became 'plutonium'; never a classicist, he had no idea that the material soon to be at the heart of the atomic bomb was named after the Roman god of the afterlife.†

Every element is given one or two letters to identify it – usually the first two letters of its name. Neptunium would be Np (Ne was taken by neon). Plutonium should have been 'Pl', but Seaborg couldn't forget how much the Gilman Hall attic had reeked. In the stinkiest joke on the periodic table, he immortalised plutonium as *pee-eew* (Pu).

Now in Chicago, Seaborg still needed a way to mass-produce plutonium. At the rate they were working, it would take 20,000 years to make a bomb: realistically, they needed

---

* This wasn't the end of code words. During the Manhattan Project, the fissile isotope of plutonium, Pu-239, became known as '49' – the last digits of 94 and 239.

† In 2006 astronomers decided Pluto would be reclassified as a 'dwarf planet', rendering Seaborg's stylistic choice pointless.

to produce a billion times more of the world's rarest substance. 'A billion is a hard-to-conceive number,' Seaborg explained in his autobiography. 'If you took a ball an eighth of an inch in diameter and increased its diameter a billion times, it would be close to the size of the Moon.' Worse, they then needed to make sure it was the fissile Pu-239; anything else would disrupt the chain reaction and ruin the bomb.

Fortunately, Seaborg had a genius on hand to help. Joining him in the Windy City was Enrico Fermi. There, on a rackets court located under the bleachers of the university's sports stadium, 'The Pope' was creating something only his fertile, febrile mind could imagine. On the hardwood floor, Fermi and his assistants had built a mound of graphite blocks, a 'crude pile of black bricks and wooden timbers'. By feeding in rods of uranium into the heap to start the neutrons bouncing around, he was hoping to create a sustained, controlled neutron-capture reaction that would produce fissile material on a scale never seen before.

It was the first nuclear reactor. But Fermi's 'Chicago pile' was never going to be permanent: nobody wanted a nuclear reactor in a sports stadium in the middle of a city. They needed somewhere quieter and less populated.

Tennessee was perfect.

★ ★ ★

'Y'all goin' up to the lab?' The barista beams at me as she slides my latte over, the spring in her step as effervescent as its foam, her rich Southern accent flowing mellifluously. I laugh.

'What gave it away?' I'm wearing my smartest (crumpled) charcoal grey suit and a loose shirt suited to Tennessee's humid climes. Knoxville is a sleepy town on the main highway, tucked in the far eastern extremes of the state. Here the Great Smoky Mountains roll like ruffled green carpet into the hazy distance, a mist-shrouded heartland that gave birth to Davy Crockett, king of the wild frontier, and Jack Daniel, king of wild whiskey-soaked nights. Rising up the steep banks of a gully carved by the Tennessee River, it could be any small town on the continent, were it not for the Sunsphere – a giant golden ball on a stick left from the 1982 World's Fair – and the university's 100,000-seater Neyland Stadium. Currently, its main claim to fame is having a former WWE wrestler as its county mayor. I could be anyone, here for the state's trinity of God, country music and American football.

The barista leans forward, tapping a nail on my reading material: *The Making of the Atomic Bomb*. 'That was kinda a clue,' she winks. Sheepishly, I tuck my book away, grab my wake-up juice and head for my ride.

It's a conversation that couldn't have happened 75 years ago. Until the Second World War, the area just west of Knoxville was little more than sleepy, wooded valleys and quiet farmsteads. The few families there were poor, often surviving on as little as $100 a year. There were no roads to speak of beyond the highway save a few dirt tracks. All it had in abundance was electricity: during the Great Depression of the 1930s, the government had tried to rectify the abject poverty with schemes to improve the area, including the construction of huge hydroelectric dams throughout the region.

These factors made it the perfect site for a secret nuclear base. The region was close enough to the highway and railroad to improve the roads and transport goods but isolated enough to avoid detection. An easy, plentiful supply of power was on the doorstep. And, most importantly, the terrain meant that the different parts of the Manhattan Project based in the area could have their own valley; if the nuclear reactor happened to blow up, the different strands of uranium enrichment and plutonium separation wouldn't all vanish, as the general in charge argued, 'like firecrackers on a string'.

At the end of 1942 the homesteaders of Scarboro, Wheat and other small villages along the Clinch River found eviction notices on their doors. They had up to six weeks – many less – to get what they needed and leave. In their place came the US army, occupying a stretch of land some 27km (17 miles) long. In nine months, the mud tracks and dense woods were transformed into a hidden world where, by the end of the war, 75,000 people would live and work. They ranged from some of the most famous scientists in the world to illiterate labourers, young women doing essential machine work and guard-post GIs frustrated at sitting out the war. The emerging population even had future fast food magnates: the assistant cafeteria manager Harland Sanders would go on to rebrand himself as 'The Colonel' and found Kentucky Fried Chicken.

The whole thing was a mess – Seaborg described it as an 'unfinished movie set' of rough-hewn roads and prefab houses. Everything ended up smeared in clay from the unpaved roads. But underneath the mud was science on a scale never witnessed before. In one valley sat the largest single building in the world, the K-25 gaseous diffusion plant, its 152,000m^2 (1,640,000ft^2) of floor space dedicated to enriching uranium. In another was Y-12, home to the calutrons, large magnets on a racetrack for separating uranium isotopes. Finally, in a humble building of corrugated iron that looked like an old steel mill, was the X-10. This was Fermi's pet project: the reactor that would solve Seaborg's woes.

The complex that had taken over the valleys was given the deliberately uninspiring title of the Clinton Engineer Works. The new settlement also needed a name, one that wouldn't raise eyebrows with spies. In the end, the army settled on Oak Ridge.

Today the heart of the secret city is Oak Ridge National Laboratory, one of the largest research facilities in the world. It's home to some of the most cutting-edge science on the planet, its mission focused on clean energy, extreme materials and ambitious projects using advanced equipment no other research centre can offer. In one lab a team has 3D-printed a full-size submersible hull. In another, they're using high-powered beams to create nanoscale circuits. A third runs the biggest carbon-fibre research facility in the US, developing space-age technologies in collaboration with commercial companies. Summit, the lab's latest supercomputer, is the fastest in the world: the size of a basketball court, it guzzles more power than the nearest city and comfortably processes some 200,000 trillion calculations a second.* One of the lab's latest facilities, the most advanced neutron source in the world, cost around $1.5 billion. We're a long way from Fermi's Roman corridors or Seaborg's stinky attic.

'About $4.5 billion has gone into the lab in the past three years,' says my guide, former associate laboratory director James Roberto, as we traverse the site in his car. Oak Ridge is more like a college campus than a research lab, an open plaza with manicured lawns and modern buildings. The only real difference is the national-level security, the ban on alcohol, and smoking is only permitted in designated areas. The facilities are available to scientists from anywhere in the world as long as they have a good research pitch. 'This behind-the-wall secret place transformed into a mostly open scientific lab,' he continues, dodging past a flock of wild turkeys by the roadside. 'We have 3,200 research guests that come and work with us every year … our mission is science and security, but to add value too. We come up with a billion-dollar innovation at least once a decade.'

It's only when you leave the main strip and get over the ridges that you start to see Oak Ridge's industrial heart. 'At the end of the war, we had the Clinton Pile, now called the Oak Ridge [X-10] Graphite Reactor,' Roberto explains. 'That was the best neutron source in the world. [Theoretical physicist] Eugene Wigner had a vision: why not build a national laboratory around it? The US was in the process of trying to build a commercial nuclear power industry, and we needed a reactor that could be used for fuels and separation chemistry … back then, we made iodine-131, phosphorus-32 and carbon-14 because nobody else could make them. Today we still make the isotopes nobody else can make.' Within a

---

* It's weird to know I've walked inside the world's fastest computer. The operators find it hard not to make jokes about movies such as *WarGames* or being able to play games such as *Crysis.* I did ask what happened if it was accidentally turned off. 'We turn it back on again,' came the puzzled reply. 'It's just a computer.'

year of the war ending, Oak Ridge was already shipping radioisotopes to hospitals in the US. This is the life-saving end of radiation.

We pull up. In front of us, nestled in the woods, is a small nuclear reactor. 'We designed and built 13 reactors. The one we're going to is the thirteenth, completed in 1965. You won't see us changing the fuel today, but …'

'Wait, we're going into a nuclear plant?'

Roberto grins and leads the way. 'Just don't break anything.'

★ ★ ★

The X-10 – the Clinton Pile – still exists. Today it's a historic landmark and museum, a chance for visitors to walk up to the reactor wall and see how things were done before safety was a priority. It still gets checked for radiation regularly, and it's still so secret visitors aren't allowed to take photos until they're inside the building. But in 1943 nobody was even sure if it would work.

The reactor was simple and elegant in design. It was a large, 7m (23ft) square chunk of graphite – the same thing you use in pencils – with 1,248 holes bored in its side. Safely behind 2m (7ft) of concrete, three men would ride an elevated platform to the holes and feed 15cm (6in) slugs of uranium as fuel, poking them through with long rods. The uranium fired off neutrons, the graphite slowed them down and the nuclear reaction started. Once a rod was used up, the team simply returned to the hole and poked in another; the used rod was pushed down a chute and into a tank of water. From there, the irradiated fuel elements would be moved through a transfer canal to an adjacent building, where the plutonium would be extracted. The whole thing even had a safety feature: several cadmium steel rods, suspended above by an electromagnet. Cadmium absorbs neutrons, so if things got out of hand all the team needed to do was kill the power and the rods would crash down to end the reaction (Fermi called it a 'scram' system – because if a nuclear reactor is going into meltdown, that's what you need to do).

Figure 3 Feeding uranium rods into the Oak Ridge X-10 Graphite Reactor.

The reactor went critical early on 4 November 1943 and the first plutonium was produced in mid-December. The project team just needed a way to separate Pu-239 from all the other bits.

Up in Chicago, Seaborg had spent a year assembling a team of chemists to tackle the problem. This was easier said than done. Most of the original element leaders – Lawrence, Fermi, Abelson, Segrè, McMillan – were busy working on problems of their own in a sprawling spiderweb of secret research that included the biggest names in physics. Seaborg was a virtual unknown in his early thirties who had to somehow persuade the best scientists in America to sacrifice their careers and work on something he couldn't tell them about. He wrote in his diary his sales pitch: 'No matter what you do with the rest of your life, nothing will be as important to the future of the world as your work on this project right now.' Seaborg's team blossomed to 50 scientists.

One of the first was Stanley Thompson, Seaborg's oldest and closest friend. Born a month apart, they had started high school together in Los Angeles and had bonded over their shared passion for science. Thompson had been rough and tumble ('the kind of guy who would tackle you without warning and start a wrestling match', Seaborg recalled), but had proven to be just as talented – if not more – than his classmate when it came to chemistry, and like Seaborg had gone on to study at the University of California, Los Angeles. Here, Thompson had shown he was willing to look after his friend when things got tough: when the impoverished Seaborg had almost dropped out of university, it had been Thompson who had loaned him the money to continue.

Thompson was working for Standard Oil, but Seaborg knew he was exactly the kind of person the plutonium project required – a 'chemist's chemist'. In his autobiography, Seaborg recounts how he soon lured him to Chicago with a letter:

> *The work here is extremely important, perhaps the number one war research project in the country, and it is of such a character that it will almost certainly have post-war significance and develop into a large industry [...] unfortunately I cannot divulge to you the nature of the work but, knowing the nature of my activities in the past, you are in a fair position to guess. It is research work of the most interesting type; it is the most interesting problem upon which I have ever worked.*

Within months, Thompson blossomed into a world-class talent. His work in petrochemicals meant he had experience in scaling up reactions unmatched by his academic peers, while his intuition, patience and attention to detail made him the lab's leading experimentalist. Seaborg later described him as 'the best experimental chemist I have ever known'. High praise from a man who knew virtually every prominent scientist of the late twentieth century.

A more maverick addition was a relative stranger. One of Helen Seaborg's friends in the secretarial pool, Wilma Belt, had met a technician called Al Ghiorso, who had been hired by Berkeley Lab to wire up the intercom. The couple had married, and Al had stuck around building Geiger counters for the lab: a tedious job, but all he could get with his modest qualifications. Ghiorso cut a strange figure. He was a full head shorter than Seaborg, with hair thick with pomade, eyes hidden behind dark-rimmed glasses and a pen permanently anchored into the top pocket of his white shirt. He had grown up in Alameda, a short distance from Berkeley, on a ranch where his father bootlegged liquor during Prohibition. Living under the flight path of Oakland Airport, the young Ghiorso had taken an interest in planes, then in radio. Constantly tinkering, by the time he was in college he had boosted his ham radio to the point where he could speak with people in Ohio, some 4,000km (2,500 miles) away. This was well beyond the world record for the 5m band; characteristically, Ghiorso was operating illegally (he had never bothered to get a radio licence), so never claimed the world record prize.

In mid-1942, as the US began to mobilise more men, Ghiorso had become concerned he'd end up drafted into the regular army, and had decided to apply for a commission in the US Navy. Lacking the required references, Wilma suggested he write to Seaborg – someone he barely knew – for a recommendation. When the letter arrived in Chicago, Glenn shared it with Helen. Realising that Ghiorso had exactly the kind of crazy, technical nous her husband needed, she looked at Glenn and said: 'You hire this guy.' Seaborg's return letter had included both the recommendation and a job offer. Ghiorso chose the latter. The two wives had conspired to create one of the most successful research collaborations in history. Unlike Thompson, Ghiorso didn't jump at the Chicago offer; he only signed up after making Seaborg swear he'd never have to make another Geiger counter again.

Ghiorso had a stubborn, non-conformist streak and thrived on being contrary. On paper, he was the least qualified of

Seaborg's team – for his entire career, he refused to get anything higher than a bachelor's degree in electrical engineering. He never owned a TV (convinced it would rot his brain), constantly doodled on anything he could find and didn't hesitate to express his radical, liberal views. He was also compassionate and caring; in later years, when one of his colleagues, Mike Nitschke, was dying of AIDS, Ghiorso took on the role of carer – physically, mentally and financially – and established a memorial fund in his friend's name. Ken Gregorich, a former senior staff scientist in Berkeley's heavy element team, describes him as 'an eccentric character … a little too enthusiastic at times … he was an inventor, not an engineer. An engineer is a real profession, and inventor is not. He just thought about the way he knew things worked and would go "well, it ought to work this way too" and he'd go and try it.' Another Berkeley alumna, Dawn Shaughnessy, also remembers his unorthodox style to lab work: 'You'd hear stories of Al breaking targets and not telling anyone. People would walk in the lab and say: "What's that cloud in the sky?" And Al would reply: "Oh, that's radioactive, don't inhale it and you'll be fine …"'

The first task for Seaborg's newly assembled team in Chicago was to work out how to get as much plutonium-239 as possible. This meant pioneering a new field called ultramicrochemistry, with its own unique apparatus. Amounts were weighed using a hair-thin quartz fibre fastened at one end; the amount the fibre bent gave you the weight of your sample. It was, Seaborg recorded, 'an invisible material being weighed with an invisible balance'. In December 1942, while Fermi played with his pile under the bleachers, Stanley Thompson hit upon a process (using bismuth phosphate) that could extract the plutonium more effectively than before. The quantities available – and the consequences of failure – meant the team couldn't afford to waste a single drop.

Even so, accidents happened. The chemists had to do these separations behind a wall of lead bricks to avoid a dose of lethal radiation, using a carefully placed mirror to guide their hands. One night a lead brick fell on a beaker, leaving around a quarter of the world's entire supply of plutonium-239 soaked in the Sunday edition of the *Chicago Tribune*. Fortunately, a little chemistry went a long way: the team reclaimed all of it by dissolving the newspaper in acid.

General lab safety was another challenge. Even though plutonium only produces alpha particles (which can be blocked by as little as a sheet of paper), if it enters the body it can end up in bones, creating a permanent source of internal radiation that can slowly and silently kill. One night, a worker using his bare hands gripped a test tube too tightly and it broke in his palm. Seaborg collected about 1mg of plutonium from his colleague's hands – the weight of a small snowflake. It was enough to force the scientist to wear gloves whenever he ate or drank until the radiation count faded.

Burnout was a danger too. The scientists worked six days a week, usually all day and into the evening. Seaborg began to have anxiety attacks at night, and the stress and nervous exhaustion finally manifested as a fever that put him in hospital. To fight off his ill mental health, the team leader began to exercise vigorously and was soon addicted to golf, climbing steep hills and running up stairs. The anxiety faded.

In Chicago, Fermi had demonstrated that a sustained nuclear chain reaction was possible, while Seaborg's team had shown that plutonium could be separated. At Oak Ridge, the two innovations had been put together. In 1944 the team was ready to scale up production. Thompson, the obvious choice to oversee the chemistry, was sent to Hanford, Washington, 1,550km^2 (600 miles^2) of plant deep in the isolation of the Pacific Northwest, to oversee the chemical arm of the first full-scale plutonium production line. Synthetic elements were about to be produced on an industrial scale.

* * *

Roberto and I have arrived at the control room of the High Flux Isotope Reactor (HFIR) at Oak Ridge. We're looking out from the gallery window at a pool, a strange blue light coming up from below the water. Deep below the surface,

5m (15ft) down, is the top of a nuclear reactor. The glow is Cherenkov radiation – charged particles passing through water to show that a fuel element is nearly spent. The brighter the blue, the sooner you need to replace your fuel. 'That one is about a month old,' says Kevin Smith, former deputy division director for the reactor research division. 'We're getting ready to shut it down.'

He doesn't mean the whole reactor – HFIR is due to keep going for decades. He just means changing out the core. HFIR's reactor fuel isn't fed into slots by three men on a lift but lowered in compact cylinders. The water is everything, cooling the reactor and slowing the neutrons. The whole tank flushes through about 275,000l (60,500 gallons) a minute, shooting through at about 15m (50ft) a second. Instead of graphite, the fuel is surrounded by beryllium – an excellent neutron reflector. 'So, neutrons will reflect, bounce off and come right back into the core,' Smith says. 'This is all a flux trap, neutrons leaking in.'

This is by design. While most nuclear power plants focus on energy, the aim here is neutron production (even so, HFIR produces 85MW of thermal power, theoretically enough to supply a city of 100,000 people). It's one of only two facilities in the world focused on making the first couple of elements beyond uranium, although it can go all the way up to element 100. The other, the Research Institute of Atomic Reactors, is in Dimitrovgrad, Russia.

Plutonium is still as much in demand as it was during the Second World War, although today's reasons are entirely peaceful. 'We're making Pu-238 for NASA,' Smith says. 'Anywhere the sun doesn't shine, you can use it to generate power. If you go to outer space, and you're far away from the sun, you got nothing else.'

'It's being used in probes?'

'Yes, *Voyager* is powered by plutonium,' Roberto adds. 'And the Mars *Curiosity* rover too.'

It's far from the only point of pride in the reactor's output. The products are used to improve the safety and efficiency of solar cells, computer hard drives and advanced medicines.

Previous reactors at the lab have even played secret, silent roles in history. In 1963 neutrons from the Oak Ridge Reactor – HFIR's predecessor – were used to bombard lead fragments from bullets found at a murder scene and gunshot residue from paraffin casts of the suspected murderer's hands and face. Using the neutrons, scientists were able to determine that the bullets all came from the same weapon – and that Lee Harvey Oswald's gun had assassinated President John F. Kennedy (the casts directly linking Oswald were inconclusive).

Standing this close to a nuclear reactor is awesome. The whole thing looks like an industrial movie set – lots of monitors, dials, gauges and unlabelled buttons that glow when you press them. Perhaps the most amazing thing is that the design was just someone's thesis. Back in the 1950s you couldn't get a nuclear engineering degree from a university, so a school was set up on site to teach nuclear engineers. Somehow, without the aid of a computer, a student called Dick Cheverton led a team that designed something that nobody's been able to improve since; now in his nineties, Cheverton is still on Oak Ridge's books. His reward was a diploma that proclaimed him to be a 'Doctor of Pile Engineering'. DOPE for short.

We walk over to a dummy cylinder (the kind loaded into the reactor) that's propped up inside the control gallery. Half a metre (2ft) long, its interior consists of three rings, like a bullseye. The outer and middle sections are filled with 540 metal plates, curved to provide constant width for the coolant to flow through. From the top, it looks a bit like a jet engine. You stick your target rods in the middle, and all the neutrons get directed back to them. Smith walks over and picks up what looks like a short aluminium javelin – or 'rabbit', as the team prefers. 'You screw off the top and then you can put seven of these target rods in there.'

He passes the target rod to me. It's basically just a hollow tube to hold your target pellets. Then it sits in the reactor for a few cycles (about 24 days each) until the pellets turn into something else. The whole thing is ridiculously light, so much so I could balance it on the end of my … *crash*.

'You've broken it,' Roberto laughs.

Smith sighs. 'That's what always happens. I'll get an operator to put it back together.'

'I hope it wasn't expensive,' I mumble sheepishly.

Another sigh. 'These little guys are about $10,000, something like that.'

Time to leave. Quickly. My next stop is up the hill, where the chemists separate out the reactor's products using similar techniques to those pioneered by Thompson 70 years earlier. They don't just produce plutonium. By December 1943, with work in Chicago slowing down, Seaborg's scientific mind had drifted away from just the Manhattan Project and back toward element hunting. In the two and a half years since its discovery, plutonium production had gone from a few atoms to quantities large enough to use in experiments. What would happen if you put plutonium in a cyclotron and bombarded it? Would another neutron be captured and turned into a proton? Could there be element 95, or even 96? Seaborg formed a small team to find out.

The answer would end up rearranging the periodic table.

CHAPTER FOUR

Superman vs the FBI

In April 1945 the New York offices of Detective Comics, Inc. received a visit from the FBI. Polite yet forceful, the agents were taken to meet publisher Harry Donenfeld. It was about the company's most popular comic strip. The G-men demanded that the latest syndicated story be pulled from publication immediately. Donenfeld called in editor Jack Schiff, who was running the storyline.

Why, the FBI agents asked Schiff, was Superman leaking state secrets?

In the latest strip, *Science and Superman,* the Kryptonian hero had agreed to undergo a few tests in a particle accelerator for a scientist. 'No, Superman! Wait! Even you can't do it!' his lab-coated ally warned, panicked by the idea of someone being hit by 'electrons travelling at 100 million miles per hour and charged with three million volts'. The equipment – and the numbers – were a little too on the money to be a coincidence. Fortunately, the FBI soon realised no one was a spy: the writer, Alvin Schwartz, had simply copied the idea of an 'atom smasher' from something in a 1935 issue of *Popular Mechanics.** The article had described one of Ernest Lawrence's cyclotrons.

A particle accelerator is basically a giant gun. Instead of bullets, it fires electrically charged particles down a vacuum tube, which contains a series of electrodes. By flipping the

---

* There are several versions of this story; I'm taking as a basis Schiff's word, although doubt has been cast on whether the strip was cut, replaced or monkeyed with in any way. In April 1948 *Harper's* published a secret memorandum from the time, which insisted that the strip was harmless and 'will considerably de-emphasise any serious consideration of the apparatus to many people'.

polarity of the electrodes at the right time, researchers can push and pull particles down the tube and make them go faster and faster. It's a carrot-and-stick approach, using the same idea that makes a TV or X-ray machine work.

The first particle accelerators were 'linear accelerators', which shoot particles in a straight line. The problem is that to get a charged particle up to the kind of speed (and therefore energy) needed to punch through a nucleus's Coulomb barrier requires an accelerator more than 100 metres long. This is far too large to fit into most labs.

Enter Lawrence's invention. A cyclotron fires particles in a spiral, starting in the centre and looping out through two giant semi-circular electrodes called dees (because of their D shape). The whole thing looks like an oversized zinc battery, sandwiched under a giant magnet that helps the particle to bend around the spiral track due to something called the Lorentz force. With every completed loop the particle gains velocity, before it finally whizzes out of the machine.

Both linear accelerators and cyclotrons have been used to discover elements. Once the ions (atoms stripped of electrons so they have an electric charge) have been accelerated, they are rushed down a 'beam line' toward whatever target the researchers are trying to hit. Then all the team can do is sit, wait and hope for the best.

Thanks to their circular shape, a cyclotron is far more compact than a linear accelerator. Lawrence's first accelerator, the size of his hand, was made out of copper tubes, wires, a vacuum pump, sealing wax and a kitchen chair: the whole thing cost about $25. By 1932 he had built a device 69cm (27in) in diameter, capable of accelerating particles up to energies of 4.8 million electron volts (MeV). This isn't much in the scheme of things – a neutron produced by fission has an energy of 2MeV – but it was more than enough on the atomic scale. Lawrence's breakthrough meant you didn't need a particle accelerator the length of a football field any more.

Not everyone was impressed: you still had to hit the nucleus, something unimaginably small. 'You see,' Albert Einstein said dismissively in 1934, 'it is like shooting birds in the dark in a country where there are only a few birds.' Einstein was right. Yet even in a dark country, you'll eventually hit something if you keep going for long enough. The nucleus had been hit before – and the cyclotron effectively gave everyone a super machine gun with an infinite supply of bullets. One of the later element creators, Mark Stoyer, explains it to schoolkids by asking them to throw marshmallows at each other's mouths from the other side of the classroom. Most of the time they miss entirely; sometimes they get an unwanted reaction and the marshmallow bounces off a nose or an ear. But sometimes – rarely – they get lucky and it goes in. 'Now,' Stoyer ends, 'imagine you're throwing 6 billion bags of marshmallows a second. For three months. And every bag has 1,000 marshmallows in it. Science gets messy sometimes.'

Lawrence had made larger and larger cyclotrons. By the time the invention won him the Nobel Prize (disturbing his tennis match and annoying his secretary in the process), his

Figure 4 Ernest Lawrence (bottom row, fourth from left) and his team sitting on the magnet of the 60-inch cyclotron, 1939. Among those on top of the machine are Phil Abelson, Luis Alvarez, Edwin McMillan and Robert Oppenheimer.

latest cyclotron was 150cm (60in) wide, its magnet large enough that the whole Berkeley Rad Lab – 46 people – took a group photo of them sitting on top of it. There were other cyclotrons scattered around the world too: James Chadwick had built one in Liverpool, and the Germans and Russians both had one, as did the Japanese. It was hardly a state secret.

Back to Superman at the Manhattan Project. Schiff had refused the FBI's request to pull the offending panels, only to be overruled by his publisher, who had a ghost writer come in and make changes. The storyline was quickly wrapped up: the hero survived his encounter with the cyclotron ('never felt better!') and the strip was quietly changed to something more all-American, where the Man of Steel played a baseball game single-handed. As *Newsweek* commented after the war, 'Superman could take [a cyclotron bombardment] and did. What he couldn't take was the Office of Censorship.'*

A year before Clark Kent's accidental espionage, Seaborg's hand-picked unit of element hunters had already started to try and make element 95 with a cyclotron. In a host of locations – including Berkeley, St Louis and Oak Ridge – the team bombarded Pu-239 with deuterons and neutrons to try and induce neutron capture. The results were all negative. Perhaps there was a problem with detection? Al Ghiorso began to come up with new, innovative instruments to try to coax out any sign of a new element.

Finally, in July 1944, as the Allied forces began to break out of Normandy following the D-Day landings, an idea broke out of Glenn Seaborg's mind. What if the chemistry was all wrong?

★ ★ ★

---

* This wasn't the last word in DC Comics' brush with the FBI. In 1983 a new villain was retroactively added to its continuity: Cyclotron. And if that wasn't enough to stick it to the Feds, DC also introduced Cyclotron's grandson Nuklon – who was later renamed Atom Smasher.

Summer in Oak Ridge is hot. The hot labs at Oak Ridge's Radiochemical Engineering Development Center (REDC) are, paradoxically, cool. The 'hot' in the name is a reference to the deadly amounts of radiation that lurk inside. Fortunately, in 50 years the Oak Ridge team have never broken containment. 'You notice the doors are getting harder and harder to open as we get deeper?' one of my guides, nuclear engineer Julie Ezold, comments. 'The walls here are 54-inch concrete, each window has three panes of leaded glass that range between 3 to 8 inches and in between each of those is a mineral oil. Even then, the mineral oil needs replacing every 5 to 10 years. The radiation just eats it away.'

Ezold is accompanied by Rose Boll. Both are carrying on Seaborg's legacy of separating out the newly formed elements. Ezold is a 26-year Oak Ridge veteran who started out studying iodine, one of the most common uranium fission products. You also find iodine in your thyroid gland, which is why fallout exposure is treated with potassium iodide – you don't want the radioactive version taking up residence in your neck. Boll came to Oak Ridge via working in a hospital as a medical technologist before going back to college and specialising in medical isotopes. There doesn't seem to be one route to Oak Ridge's hot labs, but once you get there, few choose to retire.

I took a car up to the hot labs from HFIR. The newly created elements come up from the reactor via the 'Q-ball', a massive shielded container, painted white, that is usually suspended in water above the unloading pool floor. When the products are ready, it gets loaded on a tractor trailer and taken up the hill. It's hard for anything to escape 25t of protective metal.

I'm taken deeper into the hot labs' interior. Turning a corner, suddenly we're in a long gallery, where a row of chemical technicians are staring deep into inky, oily boxes in the wall – a series of enclosed chemistry stations called 'hot cells'. Their hands are gripped to giant steel rods that vanish up into the roof. From there, the rods – manipulators controlled by flexible metal tape – reach down into the boxes with metal claws, allowing the operators to puppeteer the experiments concealed inside. It's hypnotic to watch: a cross between the brute force of a power loader from *Aliens* and Tom Cruise's graceful hand flicks in *Minority Report*. The workers don't even blink as, with a twist of their wrist and a flick of their thumb, the metal claw grabs a flask of whatever they need.

The hot labs operators work 12-hour shifts, 24 hours a day. First the aluminium is trapped in a matrix and stripped away. 'Aluminium dissolves in base, whereas the target forms a hydroxide,' explains Boll. 'That's solid. You filter it out and gather it up.' Ezold's eyes gleam. 'It's pure chemistry. Just with a bit of radiation added to it.'

I approach and look over an operator's shoulder, a big guy wearing a loose T-shirt and baseball cap. Trucker chic. How he can even see inside is amazing, let alone tell where the variety of leads, pipes, buttons and wires are all supposed to go. 'It's like spaghetti,' I mutter, staring deep into the vortex above his station.

'Yeah,' the operator, Porter Bailey, agrees, his reply given in a deep Tennessean drawl. 'It gets difficult sometimes. But we have maps, procedures.' Another whirring click and a flick of his wrist, and the robot arm on the other side of the glass comes to a halt. Even with the aluminium gone, the process isn't done. You still have to separate out the uranium, plutonium and anything else there. That means taking all the solid bits that remain and throwing them in a column of acid, where the newly made elements separate out. 'We just dissolved 32 plutonium targets,' Bailey says. *Whirr. Click.* 'We're giving it an acid digest for about 24 hours.'

From this point on, nothing is wasted. Even if it isn't the element you want, the smallest bead of it is worth more than my house. 'We've been able to get material out of the hot cells in nanogram [a millionth of a milligram] quantities,' Ezold says. 'They [the hot labs staff] do magic … it's science and an art.'

The operation isn't perfect. Sometimes vials slip. Sometimes cables break. Sometimes every atom of an element on Earth ends up in a tiny, radioactive puddle at the bottom of the hot cell. 'Your eyes go pretty big,' Bailey confesses. 'Sometimes it can be several million dollars in your hand. Depends what you're working with at the time. It's very hard to put a value on this material.' The good news is that you can recover every drop by hosing the hot cell down. The bad news is that you have to restart the extraction process from the top. For the more unstable isotopes, the ones with half-lives of only a few days, that means they're lost forever.

We head out, leaving Bailey to his work, back through progressively leaner doors, stopping to stick our hands in machines with thick metal grills to make sure we're not radioactive. A Geiger counter sits nearby, clicking away squeakily. Clean enough. We pass into the less hazardous labs, full of glove boxes, white lab coats and faint caustic aromas. There are machine shops too, ready to spool out the new elements into strips of wire. This is where medical isotopes are made, ready to ship to hospitals across the US for diagnostic tests or to treat cancers. Boll describes this part of her role as being the 'atom dishwasher', just purifying the materials. She's doing herself a disservice. Every moment she applies her skill directly saves lives.

We pause in front of one fume cupboard, its bottom coated with Teflon to recoup loses if there's a spill. There's not much in there save for a small plastic bottle about half-full. 'We have about two-thirds of the world's supply right now of thorium-229,' Boll says off-handedly, before moving on.

'At Oak Ridge?'

'Uh, no. Right there. In that bottle. You're looking at two-thirds of the world's supply.'

* * *

Thorium is a strange element. It's named after the Norse god of thunder (incidentally making it the only comic book character on the periodic table), and sits a couple of places before uranium on the periodic table at element 90, between actinium and protactinium. Today, it's under constant scrutiny as a possible source of more environmentally friendly nuclear power. If you look on a periodic table, it's part of a row that sits under the lanthanides, away from the rest of the elements. This was Glenn Seaborg's great innovation in the summer of 1944.

The periodic table is built on rules: as you look down a column, the elements are supposed to have similar properties. These are based on their electrons. As mentioned before, chemistry is all about the outer shells of electrons and elements trying to fill them. When you reach the lanthanides, their shells are so complicated they all end up reacting in basically the same way as each other, even though electrons continue to be added. This is why, rather than try to put them in the main periodic table, science had condemned them to be the weird bit at the bottom, stuck in a row of their own.

Up until Seaborg's epiphany, actinium, thorium, protactinium and uranium were all set out in the main periodic table, placed at the bottom of the area known as the transition metals. It made sense: they behaved much like everything else. But neptunium and plutonium didn't. What if, Seaborg wondered, there was a second line of elements that acted like the rare earths? This would mean that all the tests they were using to try to isolate element 95 wouldn't work – the chemical reactions wouldn't happen and the rules they had assumed wouldn't be followed. The team realised that they could have been producing the element in their tests but, because they had been looking in the wrong way, they had missed it.

By now, neutron capture wasn't enough – instead, Seaborg's team was trying to achieve fusion (combining two nuclei together to make something larger) by smashing whole elements together with enough energy to get over the Coulomb barrier, but not enough to cause fission. Starting on 8 July 1944, Berkeley's 60-in cyclotron fired helium ions (two protons, two neutrons) at a plutonium target. Once the sample arrived in Chicago, Seaborg and his three assistants – Ghiorso and a pair of chemists, Ralph James and Leon (Tom) Morgan – began to purify their sample on the new theory, using reactions they knew would separate out elements that behaved like rare earths rather than the transition metals as they had assumed. Soon strange readings were detected: alpha particles at a range never seen before. Instead of transition metals, it became clear that they were dealing with a phenomenon, like the lanthanides, that began with actinium.*

New element fever took over. It was a world of complex, multidisciplinary science; work that emerged off the back of 80-hour weeks during a hot Chicago summer. Soon, they found evidence of plutonium-238. Working backwards, they mentally added an alpha particle to it. That would make it element 96 with an atomic weight of 242. Soon after, using the chemistry of the lanthanides as a guide, the team found element 95. Seaborg's actinides were real.

The chemists immediately grasped what was happening. Ghiorso, predominantly a tinkerer, was a little slower on the uptake. '[The report] read "observed and understood by Albert Ghiorso",' he would later comment at a talk celebrating the discovery 25 years later. 'I am sure I observed it ... I am sure I didn't really understand it.'

Ghiorso wasn't alone. The actinides changed how chemists thought of the periodic table, and have been the source of never-ending arguments ever since. In 1955, for example, Seaborg used his knowledge of the actinides to predict which of element 95's electrons would form partly covalent bonds with chlorine. While this may sound technical, this is crucial to understanding how to recycle and recover spent nuclear fuel rods in reactors. Seaborg, it turns out, was right – but scientists couldn't prove it until 2017.

Another lab event Ghiorso became involved with would have an even more dramatic effect, even if no one realised it

---

* Technically speaking, both the lanthanides and actinides (officially 'lanthanoids' and 'actinoids') are transition metals too, but here it's much easier to think of them as separate entities. There's currently massive debate as to which lanthanide and actinide pair (if any) belongs on the main bit of the periodic table, and an international group of chemists are trying to sort it out.

at the time. One day, playing with plutonium, Ghiorso set up his detectors to look for evidence of fission – big 'kicks' as atoms broke apart. Soon the kicks came, appearing every 15 minutes like clockwork. Ghiorso rushed around the lab telling everyone – evidence of spontaneous fission! He was detecting newly-formed isotopes by recording them as they broke apart.

'Then I happened to notice a strange thing,' Ghiorso later wrote in the book *The Transuranium People*. 'You know, those things were just like a train.' Science is full of uncertain noises and surprises – his punctual, 15-minute fissions were a little too neat. After investigating his equipment, Ghiorso realised that all the alpha radiation flooding out of his sample was charging up a plate in his equipment; the plate was just discharging and creating a false reading. 'It was a pretty good joke on me … perhaps ever since then I have had it in for spontaneous fission.'

It was a small moment, forgotten by most almost immediately. But it had sown a doubt in Ghiorso's mind about how reliable spontaneous fission was when confirming an element had been discovered. It was a worry that would eventually cause a schism throughout the nuclear world.

★★★

In August 1945, two atomic bombs fell on Japan. The death toll was staggering, the damage beyond measure. It was a human tragedy that bought the bloodiest conflict in history to a close. All the scientists could do was put the horror of their creation to the back of their mind and keep working.

Although neither of the two new elements, 95 and 96, had any military purpose, their creation via plutonium meant they still had to be kept a secret. They were also an elusive nightmare to isolate – so much so that Morgan wanted to call them 'pandemonium' and 'delirium'. But, by the end of 1945, the team felt confident to announce their discovery. Neptunium and plutonium were no longer secrets (after a plutonium bomb had been dropped on Nagasaki, how could they be?) and Seaborg decided to reveal the new elements at the American Chemical Society's national meeting. There, in front of his fellow chemists, the great element magician, still only 33 years old, planned to announce his latest trick to the world.

Fate intervened. On 11 November 1945, Armistice Day, Seaborg was asked to come on *Quiz Kids*. This was a popular Sunday night radio show – a pure slice of wholesome Americana in which children with high IQs tried to win a $100 bond toward their education. The categories varied from spelling to nature, science to literature, all met with saccharine phrases like 'that's swell' or 'gee whiz' as the precocious minds came up with the answer. Usually the special guest was a comic, but the producers wanted someone to talk about this new, exciting thing called atomic power. Seaborg happily obliged.

Ding ding! The makers of Alka-Seltzer present the Quiz Kids: five bright youngsters ready to match wits with each other and you! Seaborg sat as the intro music played. Here, behind a row of school desks, was the next generation – the children he had been fighting to protect. To his left was Sheila, aged five; beyond her, the slightly older Bob. The giant Swedish-American cut an almost ridiculous figure at the edge of the class.

As Seaborg was the guest, the kids got to ask him a few questions. At first, they were relatively easy. Then, just as the cross-examination finished, one of the kids, Richard Williams, caught Seaborg off guard.

'Oh, and another thing,' Williams asked innocently. 'Have there been any other new elements discovered, like plutonium and neptunium?'

Seaborg could have bluffed. For five years he had helped keep the greatest secret in the world; he could easily have kept quiet about elements 95 and 96 for a few days more. There were also other answers he could have given. Teams across the world had been creating radioactive elements, filling in all of the known gaps in the periodic table. He could have mentioned how francium had been found by the French physicist Marguerite Perey shortly before war had broken out, or spoken of Emilio Segrè's technetium and astatine. Although it wouldn't be announced for another two years, Oak Ridge scientists had even discovered the last missing piece of the periodic table, element 61, which they would call promethium. The jigsaw puzzle of the periodic table had no gaps – only an edge that was Seaborg's to explore.

But the chemist couldn't resist a chance to showboat. 'Oh yes, Dick,' he replied with a toothy grin. 'Recently there have been two new elements discovered – elements with the atomic numbers 95 and 96 out of the Metallurgical Laboratory, here in Chicago. So now you'll have to tell your teachers to change the 92 elements in your schoolbook to 96 elements.'

It is the only time news of new elements was announced on a quiz show.

CHAPTER FIVE

Universitium ofium Californium Berkelium

In 1946 the scientists of the Manhattan Project scattered back to civilian life. Some, like Enrico Fermi, stayed in Chicago to found the Argonne National Laboratory; some headed to the East Coast, some along the West. Some never returned at all. In September 1945 an accident with a plutonium core had killed a promising scientist, Harry Daghlian. A few months later, the same core had gone on to kill again. Louis Slotin – the man who had assembled the Gadget in the McDonald bedroom – had been fiddling with the so-called 'demon core' in Los Alamos, the lab where the nuclear bomb was designed. He had been performing his party piece, something his colleagues described as 'tickling the tail of a sleeping dragon'. Protected by little more than jeans and cowboy boots, Slotin would lower the two halves of the core together, using a screwdriver to keep them apart and prevent a fission chain reaction. With the inevitability of a health and safety training video, the screwdriver had slipped. Slotin had been engulfed in a blue flash as the air around him ionised, and died in agony nine days later. The new elements were not toys.

Most of the Berkeley researchers – including Luis Alvarez, Edwin McMillan, Emilio Segrè and Glenn Seaborg – opted to return home. In their absence, the San Francisco Bay had moved on. The new federal prison was showing its age, with its cell house damaged by grenades and gunfire after a failed escape attempt known as the Battle of Alcatraz; past that, at the base of the Golden Gate Bridge, beachcombers had unearthed an unexploded Japanese torpedo. Life would not be the same.

The men who came home, now in their late twenties and early thirties, had changed too: they were already far beyond full professors in terms of raw experience. 'We had gone away as boys, so to speak, and came back as men,' Alvarez wrote in his autobiography *Alvarez: Adventures of a Physicist.* Lawrence set them free to work on whatever interested them. 'Ernest,' Alvarez noted, 'was always a wise scientific parent.'

Alvarez had seen more than his share of nuclear horror. He had been in a B-29 bomber high above the Trinity test as a scientific observer, and had repeated the experience over Hiroshima. His friend Lawrence Johnston, working with him, had been present at the bombing of Nagasaki, and thus was the only person to complete the nuclear hat-trick of witnessing every atomic explosion in the Second World War. Within a month, the US would drop yet more of these deadly creations, setting them off in the waters around Bikini Atoll in the Pacific. Those who witnessed the tests, lacking protection and bathed in radioactive spray, would have their life expectancy slashed by an average of three months.* Seaborg would later call it the world's first nuclear disaster.

It was something Seaborg understood all too well. As one of the leading chemists in the Manhattan Project, Seaborg had been part of the committee that had discussed how the bomb should be used. He had counselled restraint, though he never regretted the choice to use his element in an act of war: he had cousins who were stationed in the Pacific islands, dreading the inevitable invasion of Japan. 'For years after the war, at family reunions,' he recalled in his autobiography, 'they made a point of thanking me … they were convinced that the bomb had saved their lives.' Even so, Seaborg felt an almost overwhelming responsibility to control his creation.

Seaborg returned to Berkeley as head of nuclear chemistry. He was the obvious choice; the boy who had grown up with nothing was suddenly one of the most important men in

---

* If this doesn't sound like much, try slashing your life expectancy by three months and see how you like it.

America. The element maker had been tempted by an offer from Chicago ($10,000 a year, about $140,000 today and four times his pre-war salary), but the lure of home was too strong. At his side, as ever, was Helen. Ten days after their return, she gave birth to twins. 'Two fragments,' Seaborg announced to his colleagues. 'But not fission.'

Seaborg had followed up his *Quiz Kids* appearance with another radio show, *Adventures in Science*. There, he was asked what his two new elements would be called. 'Well, naming one of the fundamental substances of the universe is, of course, something that should be done only after careful thought,' he hedged. 'Naming neptunium after the planet Neptune, and plutonium after the planet Pluto, was rather logical. But so far, the astronomers haven't discovered any planets beyond.'

Listeners were asked to offer suggestions. Some wanted Latin titles such as 'proxogravum' and 'novium'; others focused on the cosmos with 'sunian', 'cosmium' and even 'bigdipperean'; yet more wanted 'washingtonium' and 'rooseveltium' after US presidents; one listener even suggested, given 96 was the offspring of another element, 'bastardium'. But Seaborg was morphing into a canny diplomat: he was going to use the names to cement his actinides on the periodic table.

The actinides were now appearing in a row below the rare earths in chemistry labs around the world. To push the association beyond all doubt, Seaborg paired them up with their rare earth equivalent or homologue (group mate). Element 63 was 'europium', after Europe – the element beneath it, 95, would be 'americium'. Element 64 was 'gadolinium', after Johan Gadolin, an eighteenth-century Finnish chemist who had discovered the element yttrium – Seaborg decided element 96 would be 'curium', after the legendary radiation pioneers Marie and Pierre Curie. It was the first element named, even in part, after a woman.

The new elements needed exploration. Fortunately, Seaborg brought with him the kernel of his Chicago team. Among them was Stanley Thompson, who had given up the oil industry and decided to complete his PhD at Berkeley under Seaborg's guidance; Al Ghiorso was back too, although this time his talent wouldn't be wasted making Geiger counters. It was a team that had been forged in a crucible, and their skills were unmatched in the world. Seaborg was the leader who could play politics and get the experiment approved; Ghiorso the mad inventor whose equipment could do it; Thompson the brilliant chemist who could prove what they had done.

It was the dawn of a golden age at Berkeley.

★★★

'So, there was this guy, Seaborg ... he was one of the most famous scientists ever. Fun fact: his name is an anagram of "Go Bears!".' The tour guide chirps away cheerily as she takes the new university intake past Gilman Hall, full of pep and team spirit. Science and the Golden Bears – Berkeley in a nutshell.

I'm back on campus. The chill of the Bay Area mist still permeates the air. I still wish I'd picked up one of those cheap 'I heart San Francisco' sweaters. Fortunately, this time I have a stiff climb to keep me warm. Berkeley Lab sits on the slopes of the hills above campus. At its top is a massive dome, today housing the Advanced Light Source, shooting bright beams of X-rays around in a circle to understand how the universe works. It was a sight that would have greeted the Berkeley boys on their return to the lab: while they were off in Chicago, Ernest Lawrence had been busy making ever-larger machines. In 1944, under the dome, he finished a 470cm (184in) cyclotron, eclipsing the 150cm (60in) machine used to create the new elements. This was never used for war work, but the director of the Manhattan Project, General Leslie Groves, still chipped in $170,000 (about $3 million today).

Groves' donation wasn't entirely altruistic: Uncle Sam wanted Berkeley. In November 1946 the Atomic Energy Commission took over the lab. Today it's owned by the Department of Energy. Everything the lab does has to serve the taxpayer. As with Oak Ridge, it is one of the jewels in the

US government's research crown, with an annual budget of $800 million and 4,000 staff. It's the home of smart windows that respond to changes in sunlight, the antiproton and the measurements of how fast the universe expands.

The climb is 30m (100ft) up from Blackberry Canyon to the offices at the top of the rise; a little to the south, the university students know the area as 'Tightwad Hill' – overlooking the university stadium, it's the perfect spot to watch Golden Bears games for free. Fortunately, I don't have to go anywhere near that far. Making the climb, passing the checkpoint, I pant and wheeze my way to the first building on Cyclotron Road. This is the home to Berkeley's current element workhorse, the slightly more modest 88-inch cyclotron. Here, in an office filled with far too many computer screens, journal papers in piles and general scientific detritus, I'm met by my laid-back guide to Berkeley's secrets. Jacklyn Gates is my kind of scientist: she reclines back in a black hoodie, her sleeves rolled up to reveal an intricate network of tattoos on her arms, the baggy top paired with jeans and comfortable cowboy boots. Gates is tired; she's just come off a week pulling all-nighters, working with the small team at Berkeley to ready their machine for a new set of experiments. Once the thriving epicentre of element discovery, Berkeley's heavy element team is the smallest it's been since Ernest Lawrence created it.

Gates self-identifies as Berkeley through and through. She stumbled into nuclear science by applying for a position at Argonne, just outside Chicago – the perfect place for a young researcher to start their career. She hadn't read the application quite right: the position was for Argonne-West, an offshoot located in the big emptiness of eastern Idaho. Hooked on the science but less enamoured with the location, she moved to Berkeley as a grad student.

I mention how a vertiginous slope is an odd place for a lab.

'Yeah, they originally started down on campus,' she says. 'Then they decided maybe doing all these things that cause radiation isn't a good idea in the middle of a university. "Let's move it near campus but not where you can irradiate students ...'' It's a steep hill, but the lab owned a bunch of property on it so Lawrence, when he was building one of his cyclotrons, started up here.'

That was a long time ago. 'The 88-inch cyclotron came after Lawrence,' Gates continues. 'It started running in 1962. There's also a small medical cyclotron. It can produce oxygens, carbons, fluorines ...' Today, Berkeley's cyclotron isn't set up for element discovery – although the team is one of the forerunners in the world for confirming the experiments of others. 'Discovering a new element is not easy,' Gates says – the understatement of the century. 'We've improved our techniques [since Seaborg], but we're still not perfect. These days you need a separator. We've got one, but it's not ideal for determining new elements. For what we use it for it's great. It's got high efficiency, good suppression of background, but you have to know where the products are going to come from: if you're 5 per cent out where you're setting your magnets, you won't see anything. These superheavies, the elements we're discovering today, are 6 per cent off where the projections are.'

That, Gates explains, is the real problem with element discovery – not hoping for fusion but spotting it when it does occur. 'It reduces our efficiency for detection from 60–70 per cent to 10–12 per cent. We had that problem confirming the [element] 112 experiment. We thought we knew where it would be, and we were wrong. We didn't see anything for a month of beam time. And that costs $50,000 a day. So, we spent $1.5 million because our magnets were 6 per cent off, and there was no way to know that.'

The expense and difficulty mean modern-day Berkeley sets its priorities elsewhere. The team simply doesn't have the money to go fishing for elements. 'The state of science funding in the US has been pretty sad for the past couple of decades,' Gates says. 'We used to put a lot of effort into basic science, things that won't get you a return on investment now but will lead to major discoveries in 30 or 40 years. We don't see that as a priority [today]. And you have to ask: is making a new element more important than climate change, or renewable energy? It's not an easy argument to make. I mean, *I'm* a heavy element person, and *I'd* pick renewable energy.'

Gates makes an important point. Although this is the story of the big stuff – the major discoveries – it's a tiny fraction of what any heavy element team does. They're not there solely for the glory of discovering a new element, as cool as that may be; every isotope unlocks a little more of the universe or creates options for saving lives through medical isotopes or cleaner energy. Only a tiny fraction of Gates's energies are dedicated to making elements. The rest of her work is fundamental science. Berkeley's heavy element team is small, but it's focused.

Gates wrinkles her nose. 'You want to get something to eat?' We head back down the hill, through the brisk air, toward the nearest bar that does sticky finger foods. I ask if we have to climb back again. 'Seaborg would have walked,' Gates teases. 'He had the lab build a set of stairs up the hill for him. They called them the Seaborg Steps. He walked them every morning for exercise, up and down, up and down. For years, every researcher knew that if they wanted his help, they could find him at the stairs. They just had to be willing to walk with him.'

I shiver a little on the way to the bar. 'You can always tell a tourist in San Francisco,' Gates comments, tugging her hoodie sleeves down, safe under the security of her warm layers. 'They think California is going to be sunny. They end up having to buy a sweater. You always see people in "I heart San Francisco" tops because they didn't bring anything …'

I keep my head down and mouth shut as we walk down Cyclotron Road.

★ ★ ★

While Berkeley Lab must justify everything it does today, the 1940s were a paradise for the physics of the unknown. A word from Lawrence or Seaborg and a project became reality. Nobody knew what the new elements could do, but they knew the awesome power of plutonium. Money was no object, and space and time were freely given. Perhaps the greatest example came a decade later, with the construction of the bevatron: a particle accelerator so vast it was the size of a modern cycling velodrome. Its name was a portmanteau of its purpose: the 'billions of electron volts synchrotron'.* Lawrence's Big Science had become bigger than he ever could have imagined.

But despite this ready source of funds, things weren't easy for the new heavy element group. For three years, the team worked building their new lab in Berkeley's Building 5, a small, cramped and crude shack near Lawrence's dome. They had a lot of problems to solve – not least the lack of equipment. 'The cupboard was bare,' Thompson would later recall during a symposium in January 1975. As he worked, Thompson finished his PhD, laying claim to arguably the greatest PhD dissertation in history: his doctorate was awarded for exploring americium and curium. But no one – Seaborg, Thompson, Ghiorso or Lawrence – was prepared to rest on their laurels. There were more elements to come.

The set-up to hunt for the next elements in sequence, 97 and 98, was relatively simple. Helium ions – alpha particles by any other name – could be fired at americium and curium, creating new atoms two places along the periodic table. But the targets posed a problem. The world's supply of americium could fit into the eye of a needle, and nobody had isolated curium at all. There wasn't anything to shoot at.

For four years, Thompson and a host of other chemists worked on how to get more. 'We worked on closed cycles and calculated masses, energies and half-lives,' he recalled. 'We used systematics, alpha half-life energy relationships for isotopes of different elements, and even developed some crude electron-capture schematics.'

In short: it was hard. Ghiorso summed up the team's efforts during the 1975 symposium: 'They would work very, very,

* A synchrotron is a particle accelerator shaped like a ring, rather than a spiral like a cyclotron. The most famous example is the Large Hadron Collider at CERN. These are amazing machines, but too powerful to create elements.

very hard with a tremendous number of separations, very difficult procedures, and end up with a small sample. Then they would hand it over to me and say "Here, we're tired, you find out what's in it."'*

The answer was americium and curium, both in large enough amounts to make targets. On 19 December 1949 the team had a hit: a target containing a mere 7mg of americium was struck with helium ions and produced element 97. As Seaborg watched Thompson work, his heart rate began to rise, pounding against his chest in furious anger. The university doctor was summoned and soon became convinced Seaborg was having a heart attack. The chemist was rushed to hospital, where he was kept under observation for several days. It soon emerged nothing was wrong: he was just too excited about his latest discovery.

Rather than wait for chemical confirmation, the creators named their new element immediately: 'berkelium'. It had been such an arse to create that Thompson and Ghiorso wanted its symbol to be 'Bm'. Seaborg overruled them and chose Bk.

The team were on a roll. On 9 February 1950 an even smaller target, 8μg of curium, produced just 5,000 atoms of element 98. The amount was invisible to the human eye, but the team were so good even this was well within their ability to handle. 'There were no false lunges at element 98,' Seaborg recalled. 'The predictions of both its radioactive and chemical properties were made with uncanny accuracy and led us to our prey without a single misstep.'

Joining the trio of discoverers was Kenneth Street, a former US marine and one of the staff who had helped Thompson during his long years of chemical toil. The four men decided the new element was going to be called 'californium'.

---

* Thompson's work ethic was staggering. At one point, he and another researcher, Burris Cunningham, worked 36 hours straight in the lab. Stepping outside, they realised Thompson had misplaced his coat. The exhausted duo spent an age searching around until one of them noticed Cunningham had accidentally put it on over his own.

When the papers officially announcing the elements emerged back-to-back in the *Physical Review*, reaction was muted. A few people objected to the spelling of berkelium, wanting to lose the second *e*; to this day, how it's pronounced (*berk-el-i-ium* or *berk-lium*) varies around the world. Seaborg excitedly phoned the mayor of Berkeley to tell him a new element was named after his town, only to have the official hang up on him in utter disinterest. Two scientists from the Soviet Union contested the discoveries, insisting that they had predicted the properties of element 97 two years earlier based on the periodic table (wisely, the international community decided that predicting an element isn't the same as actually creating it). *The New Yorker* even mocked the discoveries, pointing out that, given they had created elements 95 to 98, the team could have written *universitium, ofium, californium, berkelium* across the bottom of the periodic table.

Reaction to the new elements was a little more positive in Sweden. The Royal Swedish Academy of Sciences decided to give Edwin McMillan and Glenn Seaborg the Nobel Prize.

★★★

The Nobels are considered the pinnacle of scientific attainment. Originally, they were supposed to be awarded to whoever made the greatest contribution to science in the previous year. Today, the awards are more commonly recognition for work in a ground-breaking field – a scientific lifetime achievement award. For the science prizes, the committees take their time, waiting to see how discoveries will pan out and whether they continue to deliver a 'richness of consequences' for humanity. No more cancer-causing parasitic worms or lobotomies.

The award process is carefully regimented. First, around 3,000 people are invited to make their nominations – secret ballots, the records of which are then sealed for 50 years. Usually, this results in around 300 potential prize winners – it's virtually unheard of for someone to win the first time they are nominated. Next, a committee goes through the possibilities, consulting even further with experts and former laureates. They investigate their potential winners in complete secrecy, digging deep to make sure the nominated person really did the work and didn't just copy it from their subordinates (if that sounds outrageous, such skeletons have been uncovered in the past). Finally, each year between one and three people are awarded the prize for a single body of work; sometimes the prize is split evenly, sometimes one person gets half and the other two get a quarter. It really boils down to who did what.

I've spoken to Nobel Prize winners about what comes next. Most do not expect it; when they receive the phone call from Stockholm, typically half an hour before the prize is announced to the world, they almost always assume it's a practical joke. The committee chair keeps someone the person knows on standby, ready to come on the line and tell them it's real. Once informed, the Nobel winner is in a strange, surreal limbo. They aren't allowed to tell anyone, so most just try and get on with work, or perhaps phone their family and hint they should watch the news. Sometimes, during the lull, they get phone calls from previous Nobel laureates. 'Congratulations,' the speaker usually says, 'this is the last 20 minutes of peace and quiet you'll get for the rest of your life.'

When the call came in November 1951, Seaborg had been forewarned. A guest lecturer had accidentally told Helen she would soon visit Sweden, and Glenn had heard speculation on the radio as he drove to work whether he would win the prize. Even so, as he took the call Seaborg was overjoyed. With the prize came instant fame, a large gold medal and a windfall of $16,000 ($158,000 today). It was enough for him to completely remodel the family home.

Seaborg delighted in returning to the country his ancestors had left behind. Arriving with McMillan, Lawrence and their wives, he found himself caught up in a world where he had to crown pageant princesses, sign autographs for delirious children and watch as flags were flown in his honour. At the Nobel banquet, he was required to answer the toast given by

King Gustaf VI Adolf. Rising, Seaborg dusted off the Swedish he had spoken at home as a child in Ishpeming. He had been practising for weeks. '*Ers Majestät, Era Kungliga Högheter, mina damer och herrar …*'

There was a shocked gasp among the audience, as if he had just sworn at the king. Seaborg couldn't understand what he'd done – he'd chosen his words with care. It was only in the morning, when he read the papers, that he found his answer. Seaborg's Swedish had been flawless, but it had never occurred to him that his impoverished, machinist parents might have had thick working-class accents.

Seaborg and McMillan did not win their Nobel for discovering elements beyond uranium (the prize is never rescinded, and Enrico Fermi had already claimed that honour). Instead, they won for 'their discoveries in the chemistry of the transuranium elements'. It was a compromise that worked for them both. Congratulations flooded in from around the world, along with a fair lump of sour grapes. 'I suppose if you can't find new elements,' quipped the British physicist Lord Cherwell, 'you just have to make them.'*

Nuclear power, and the elements that had come with it, captured the imagination of the world. Dreamers imagined cars and homes powered by quantities of uranium the size of a sugar cube. Plans emerged to make Antarctica habitable, prevent earthquakes, free up natural gas supplies and even change weather patterns. Ideas were even touted for nuclear bombs contained in hand grenades – until someone pointed out that throwing such a device would be suicide. More realistic uses became apparent too: weapons, energy and cancer treatments. In his later book *Man and Atom*, Seaborg listed 60 radionuclide sources used in medicine in 1966, treating a combined total of 33,743 patients in the US. Today,

* Cherwell was being snooty. Thanks to the likes of Humphry Davy, the British had discovered more elements than anyone else by cracking open rocks, zapping things with electrolysis or probing the air. Now the Americans were manufacturing them beyond the supposed boundaries of the periodic table. It just wasn't cricket.

nuclear medicine is a standard part of every major hospital in the world.

The synthesised elements became household names. Plutonium had given the world the atomic bomb. In time, the elements further along the table would find uses too. Americium, it turned out, released a steady stream of alpha particles, easily blocked by anything in the air. Today, my home smoke detector – and others like it around the world – contains 0.9μg of americium-241. It's more expensive, gram for gram, than gold, and probably the only radioisotope you can buy in your local supermarket.* Curium, meanwhile, is used to produce alpha particles for X-ray spectrometers in space probes, including the rovers currently roaming the surface of Mars. Berkelium and californium did not have any apparent uses, but as a demonstration of the superiority of American science they were unrivalled.

The search for elements had started as scientific curiosity. It then morphed into a wartime imperative and, as the 1950s began, changed once more into a matter of national prestige. It was a sensation that would only grow as the rival ideologies of capitalism and communism began to split the world. Already, tensions were mounting. At Berkeley, faculty members were ordered to take a loyalty oath: just one symptom of the paranoid searches taking hold across the US for anyone considered 'un-American'. In Europe, the breakdown between the West and the Soviet Union had split Germany in two. In Asia, the Korean War between the communist North and republican South, both sides backed by superpowers, had ground to a stalemate.

The Cold War had begun. And the edge of the periodic table was about to turn into a battlefield.

---

* Although the smoke detector is americium's great contribution to modern living, actinide chemists often joke that they are constantly looking for new ways to 'make americium great again'.

CHAPTER SIX

The Death of Jimmy Robinson

The F-84 Thunderjet roared through the skies, the waters of the Pacific Ocean a smooth cerulean carpet stretching as far as the eye could see. Captain Jimmy Robinson swallowed hard, his own breath echoing inside the respirator, and steeled himself for the mission to come. The next few minutes would involve the most taxing manoeuvre of his career. Aged 28 years, Robinson was a seasoned pilot who had flown as bombardier in Liberator bombers during the Second World War. Even so, he knew there was little in his experience he could call upon for the task ahead. He was going to fly inside an atomic mushroom cloud.

The F-84 was the US Air Force's main fighter-bomber. A single seater, the body was shaped like a silver clipped cigar, with straight, flat wings ending with two smaller cigars on the tips. Early designs had been a nightmare to fly (the world was still transitioning from propellers to jets), but by 1951 the F-84 was a trusted part of America's arsenal in the sky. It dominated the battlefields of the Korean War, its payloads accounting for about 60 per cent of all ground targets destroyed during the conflict. Although they were no match for the Soviet-built MiG-15s (those were left to the nimbler F-86 Sabres), F-84s would go on to fly over 86,000 missions, dropping some 55,000t of explosives. It sounds impressive until you know that Thunderjets can also drop nuclear weapons: a lone fighter could have delivered the equivalent of the entire war's bombing campaign in a single run.

Robinson was Red 4, part of a key element in a mission code-named Operation Ivy – the eighth series of US nuclear tests. Somewhere below his flightpath was Elugelab, a small islet that formed the northern bend of a 40-island horseshoe called Enewetak Atoll, today part of the Marshall Islands.

There, at precisely 07.15 a.m. local time, 1 November 1952, the US detonated 'Mike'. It was the world's first thermonuclear bomb, capable of an explosion so vicious it would mimic the intense furnace of the Sun. Mike was the next generation of atomic design, the brainchild of Manhattan Project alumni Edward Teller and Stanislaw Ulam. Rather than just imploding like Trinity, the bomb had a second stage to feed more fuel and boost the chain reaction. As it used deuterium and tritium, heavy isotopes with only one proton, the world would come to know it as a 'hydrogen bomb'.

The US Air Force had discovered by chance that you could fly in a mushroom cloud: in May 1948 a B-29 bomber observing a nuclear test had found itself unable to avoid a finger-like spur kicked up by the explosion. The pilot, Lieutenant Colonel Paul Fackler, had cut through the atomic gloom, then swung into a few rain clouds to wash down his bird on the way home. 'None of us keeled over dead and no one got sick,' Fackler had reported smugly on his return. It was unclear if it had been an accident or a deliberate stunt, but Fackler had enjoyed the experience and requested permission to form a new squadron dedicated to repeating his feat, this time equipped with scientific instruments to take samples. There's a saying in the military that the most dangerous thing you can do is give a superior officer a bright idea. *Let's make pilots fly through nuclear blasts on a regular basis for science.* While Fackler petitioned the Pentagon, someone figured they'd just get fighter pilots to fill his flight boots.

According to Paul Guthals, one of Los Alamos's cloud sampling experts, it was hard to find pilots with the 'right stuff'. In addition to flying the jet fighter – a machine that could kill you in seconds if you stopped paying attention – the selected airmen also had to run a suite of scientific equipment and recording devices. Three radiation instruments had to be monitored simultaneously, and their readings had to be both recorded on a sheet and read off to a scientist in the control aircraft outside the cloud. Each pilot also carried a stopwatch to time their duration in the cloud – and therefore their probable dose of radiation. This would have been difficult enough in clear skies, but they were being asked to do it at the heart of a radioactive dust storm. 'Most pilots with less experience and ability were simply overwhelmed,' Guthals noted in an article for the *Newsletter for America's Atomic Veterans*, 'and were often distracted by the awesomeness of the clouds' ever-changing interior.' Robinson had the guts and cool head in spades. During the Second World War he had been shot down over Bulgaria and forced to bail out, eventually ending up a prisoner of war in Romania; while parachuting down he had checked his maps, then calmly lit a cigarette as he'd waited to land. In April 1952 he had completed a practice sampling mission and proven a capable pilot. Even so, the hydrogen bomb was expected to be another prospect entirely.

At 07.15 a.m. on 1 November 1952, Robinson was still on the tarmac at base with the other pilots. There, he watched a bright orange bloom, a perfect semicircle, push out from its epicentre to bathe the sky, lightning crackling around its flame. In a lingering visual scream, the light faded, leaving only the telltale mushroom cloud of dirt, sand and coral. The Ivy Mike explosion was beyond imagination, a looming hellscape painted throbbing red from nitrogen dioxide and iron oxides. At its base, Elugelab was gone: the islet had literally been blown off the map.

Ninety minutes after detonation, Red Flight – the first of three flights that would go into the cloud that day – made their approach. The first two F-84s, led by Lieutenant Colonel Virgil Meroney, made their turn and flew inside. They were supposed to enter the mushroom's top, but at 17,000m (55,000ft) it was too high for their planes; an F-84's ceiling was only 12,000m (40,000ft). Instead, their only option was to fly into the stem, where the buffeting winds would be harshest and full of Elugelab's remains. Five minutes later the first pair of fighters burst out, jet engines howling as they zoomed clear.

By the time Robinson's turn came, the mushroom cloud had plunged the atoll into an eerie darkness. In formation with his wingman, Bob Hagan, the two fighter pilots began their approach toward the cloud. One moment it was smooth sky, the next an ugly maelstrom. Meroney had radioed back that it was filled with a 'red glow, like the inside of a red-hot furnace' and that his instruments 'went around like the sweep second hand on a watch'; Hagan described it as 'both grey and dark shades and still appears to be boiling'. Robinson grappled against the turbulence of the mushroom cloud's interior, his hands tightening on the yoke as the entire cockpit shook wildly in the crosswinds. The Thunderjet was being tossed about like a ragdoll in a washing machine, but gradually he was able to wrest back control and activate the autopilot.

Inside the bulging wing tips of his F-84, filters had been set up to scoop up whatever particles might have been created in the blast. The heart of the explosion had been a near-instant flux of 10^{24} neutrons per cm^2. Neutron capture was happening at a rate never seen before on Earth. Hungry nuclei were grabbing neutrons and remaining stable, forming isotopes as rich as uranium-255, 17 neutrons more than its most common variant, then beta-decaying into elements usually only present in merging neutron stars.

Robinson had more pressing concerns. His instruments told him the sky in front was a no-go zone, a cauldron of hot gas and broken reef. Robinson twisted his fighter out of its path, forcing the autopilot to disengage. The dust cloud was clogging up the machinery and causing his fighter to shake itself apart. The fighter's engines choked, coughed and stalled, the guttural growl flooding Robinson's senses as his hunk of metal fell from the heavens. He lost altitude, lost control, fought against the g-force not to lose consciousness. Breathing became heavy, the noise of each lungful resounding over the radio as his finger pressed into the radio button. His sinews wrestled against the vibrations and resistance even as cold sweat soaked his brow …

The fighter recovered. Robinson pulled on the yoke, levelling the F-84 at 6,000m (20,000ft). Meroney ordered him to exit the cloud. Shortly after, Hagan joined him, once again in the vast emptiness of the Pacific. The churning hell was gone; the world was serene, colourful, endless blue. It had been an impressive piece of flying.

An F-84 uses around 540kg (1,200lb) of fuel an hour. With atmospheric filters strapped to the wings, the sampler pilots hadn't been able to carry reserve tanks, so had been told to head for a tanker aircraft to refuel. If they couldn't find the tanker, they had been ordered to head south, back to the strip at Enewetak (the largest island in the atoll, at the eastern tip of the horseshoe) and land.

When Robinson and Hagen emerged from the cloud, they each had around 450kg (1,000lb) of fuel remaining. It was then they realised the electromagnetic pulse inside the cloud had scrambled their electronics and neither pilot could find the radio beacons to guide them home. They were lost and low on fuel and could see only the deep waters below. By the time Hagen managed to pick up the beacon, they were down to 270kg (600lb) of fuel and were 154km (96 miles) north of salvation.

The duo's luck worsened. Rain squalls descended and limited visibility. By the time they located the field, their machines were running on vapours. 'My gas gauge was on empty,' Hagen recalled. 'I was able to set up a pattern and land without fuel, using total deadstick manoeuvres.' Hagen was underplaying his skill: landing after a flame-out took almost superhuman effort. He hit the tarmac hard, his tyre bursting on impact. It was messy but it worked.

Enewetak was too far for Robinson. While still at 5,800m (19,000ft) he ran out of fuel; at 4,000m (13,000ft) his engine flamed out; at 1,500m (5,000ft) it was clear he wasn't going to make the field. A rescue chopper had scrambled to pick him up, and Robinson weighed up his options. Bailing out with his parachute was possible, but his flight suit included a lead-lined vest, ironically as a safety precaution to limit radiation exposure. Trying for a water landing was equally dangerous, as an F-84 wasn't designed for a life on the ocean.

'I have the helicopter in sight and am bailing out,' he reported, popping the canopy. They were his last words.

Rather than abandon his plane, Robinson seems to have thought twice about the parachute jump and opted for a water landing instead. The emergency helicopter crew watched his plane dip its belly in the ocean, skip like a stone on a pond, hit a wave and flip. They made it to the crash site, about 5.5km (3.5 miles) north of the runway, just as the Thunderjet slipped beneath the surface. 'The people in the tower told me that an airplane had just gone into the ocean behind me, and they didn't see any signs of a parachute or anything else,' Hagan remembered in the article for *Newsletter for America's Atomic Veterans*. 'Then, deep in my gut, I had a bad sinking feeling.' All that remained on the surface was an oil slick, one glove and the waterlogged remains of several maps.

Jimmy Priestly Robinson was the first person killed in the hunt for an element. Sadly, his body was never found. His colleagues expected as much: the water around Enewetak was deep, the chances of survival slim. A year later he was recognised as killed in action and posthumously awarded the Distinguished Flying Cross. By then, Fackler had persuaded the Pentagon to commission the 4926th Test Squadron (Sampling). The US would continue to fly planes into mushroom clouds until 1962.

The Robinson family had to wait longer for answers. The secrecy of atomic work meant details about Robinson's death fell into a bureaucratic black hole, with none of the US agencies willing to take up his case. His daughter became involved with veterans' groups whose dwindling networks fought to keep his memory alive. Even so, it took 50 years before a memorial was erected in his name at Arlington National Cemetery. There, in 2002, his wife was finally given a folded flag in honour of his service.

'You hear a lot about heroes,' Robinson had told the Memphis Lion's Club on his return from the Second World War. 'I don't believe in heroes.'

I do. Jimmy Robinson was one.

★ ★ ★

Robinson's death was not in vain. The other three fighters had survived the hell of the mushroom cloud, as had all of the flights that took off later, and their wing filters were rich with radioactive samples. In a nuclear blast, lighter particles fly to the top, heavier ones to the bottom. Unbeknown to the pilots, by flying through the stem, Red Flight's filters had captured something never seen on Earth.

The next steps were taken with meticulous care. Protocol at the time meant the pilots couldn't touch the outer skin of their plane, so they had to wait for the ground crew to bring out a cherry picker. Then, while the flyboys stripped down and headed to the decontamination showers, five men (known as the 'decon-grunts') would use 3m (10ft) poles to open the wing pods, remove the filters and deposit them in lead-lined containers. Sealed and secret, the cargo would make its way back to the US as quickly as possible – the samples' half-lives created a tight deadline.

Once stateside, the filters went directly to Los Alamos, the lab in charge of analysing all nuclear debris. The standard practice for nuclear particles was to dissolve them in acid, but with Ivy Mike the coral from the reef made that tricky: the samples had a habit of catching fire. Tents were quickly erected outside the main building for the precarious work. Here, it soon became apparent that something special had happened. The filters contained something emitting alpha particles with energy levels much higher than any known isotope of plutonium.

While the Los Alamos scientists got to work, some of the samples were sent on to Argonne and others to a new lab just outside San Francisco. Teller, with the support of Ernest Lawrence, had argued that concentrating all US nuclear weapons research in a single lab (Los Alamos) was unwise – and that a second lab would create some healthy competition. As an alternative, they had pitched a new facility in Livermore, a small town amid the golden hills Glenn and Helen Seaborg had driven through a decade earlier. It had been a perfect space to build an expansion of the Berkeley lab, and Teller had taken up residence there as bomb-maker in chief. Soon, the different teams confirmed Los Alamos's findings and traced the 'mysterious alpha' emissions down to new isotopes, including Pu-244. This, everyone was amazed to find, had a half-life of 80 million years – meaning it was stable enough for trace amounts to exist naturally from when the Earth formed.

The element team at Berkeley had nothing to do with Ivy Mike or the secretive work at Livermore. But winning a Nobel Prize gives you a little clout: Seaborg was a big shot, important enough to receive a secret teletype about the Ivy Mike detonation. The message mentioned that 'data on the recent Eniwetok test indicates the presence of some unique heavy element isotopes such as Pu-244'. Seaborg understood what that meant immediately. The Ivy Mike explosion had acted like a nuclear cauldron, a particle accelerator and a nuclear reactor all rolled into one. If Pu-244 had been made, heavier elements were likely. Perhaps even elements beyond californium.

Seaborg passed the news on and Al Ghiorso and Stanley Thompson made some rough calculations. The only way scientists at Los Alamos could have detected Pu-244 was by using a mass spectrograph. At the time, the best instruments were only sensitive enough to pick up Pu-244 if it made up about 0.1 per cent of a sample. In heavy element terms, that wasn't gold dust: it was a veritable *gold mine*. There *had* to be traces of heavier elements present.

Ghiorso and Thompson phoned Livermore. Kenneth Street, who had helped with californium, was there working with Teller. Calling in a favour, the pair convinced Street to hand them half of his filter paper. Seaborg was sceptical about the chance of discovering anything, but Ghiorso – just turned 37 – still felt young 'and [was] not about to be deterred by what seemed like the impossible'.

Taking their precious relic of the Ivy Mike blast, Berkeley Lab began their tests. Within minutes it was obvious that they had something that looked like element 100. Soon, they realised their mistake – it was element 99. From the time Seaborg had heard about the filter, to the discovery of a new element, had taken nine days. The team were dealing in the smallest quantities anywhere in the world. Californium, element 98, had been detected from 5,000 atoms. Element 99 was discovered from around 200 atoms. The team were quite literally plucking their prizes out of thin air.

The discovery didn't sit well with other labs, who felt they deserved equal credit. Over Christmas 1952 the entire US nuclear community began to turn on each other. Berkeley asked Argonne for more material; instead, they got a memo from Argonne claiming the lab had discovered element 100 from their Ivy Mike sample. On 15 January 1953 Berkeley also found what they believed was element 100 from the filter. The two labs began to argue.

The politics of science can get messy – and fewer things are more hotly contested than who did what first. By February, Ghiorso wrote of being 'tired of playing games with the Argonne crowd'. But Berkeley's trump card was Seaborg: the Michigan chemist had matured into a masterful diplomat. He announced Berkeley's sighting of element 100 to Argonne, but refused to tell them any specifics – preventing Argonne from adopting his team's techniques. In the meantime, he sweet-talked Los Alamos into being recognised as co-discoverers of the elements with Berkeley. Argonne's data soon turned out to be the exact error Berkeley had made: a misidentification of element 99. Seaborg's quick-footedness had outfoxed the Midwest team.

The problem was how to reveal the new elements. The Ivy Mike test results were classified; as with plutonium, the discovery couldn't be announced publicly. Instead, Ghiorso and Thompson rolled up their sleeves. If you could make elements 99 and 100 in a bomb, maybe you could make them in a lab too? All they needed was to do it before someone else.

They didn't. By the time they had produced both elements themselves, in 1954, a small team from the Nobel Institute of

Physics in Stockholm, Sweden had contacted Seaborg to tell him they had created element 100. It was a blessing in disguise: if the elements existed in the lab in another country, there was no need to keep the US findings from the Ivy Mike blast a secret. Ten days after the Swedish team announced their discovery, Ghiorso's own results appeared, taking care to mention 'unpublished [classified] information' to remove all doubt about who got there first. The Ivy Mike findings were finally declassified a year later, and elements 99 and 100 joined the periodic table.

There was still the matter of the dispute between the US labs. After a face-to-face meeting in Chicago, the discovery was given to Berkeley (99 with the aid of Los Alamos). That night Seaborg wrote in his journal that he and Ghiorso enjoyed 'an abundance of cocktails', so much so they didn't even remember flying home.

The elements needed names. Bizarrely, they already seemed to have them: a talk by Luis Alvarez had been misinterpreted a few years earlier, and some textbooks already listed 99 and 100 as 'athenium' and 'centurium' respectively. The first kooks of the nuclear age had also emerged, with one laying claim to the elements in the *Physical Review* as 'ninetynineum and centinium ... I value and honor each atom at a million dollars'.

Berkeley got to have the final say. Ghiorso made it strongly known that he wanted to name them after prominent scientists, and thought there were two obvious candidates: Albert Einstein and Enrico Fermi. Seaborg and the rest of the team agreed. Fermi was dying of stomach cancer, but Seaborg knew Emilio Segrè was in regular contact with his mentor and asked him to pass on the group decision. Segrè, blasé as ever, replied 'that he was not interested enough to do so'. By the time einsteinium and fermium became official, in August 1955, both elements' namesakes had died. Ghiorso had already written to Laura Fermi that element 100 would be named after her husband: 'It was my good fortune and privilege to know your husband [...] I can say from personal contact that science has lost a very warm-hearted human being as well as its greatest physicist.'

★ ★ ★

'Fermium' was a fitting choice for element 100. The Italian maverick, frolicking with the sheer joy of science in a fish pond, piping in radioactive gas from basement safes and building nuclear reactors under sports stadiums, had brought in the atomic age; the element named after him would bring the age of atomic discovery to a close. With fermium, the nuclei become too unstable, with half-lives too short to make in large enough quantities. Fermium atoms are too large to beta decay, so neutron capture stops being viable. Worse, later research in the 1970s showed that fermium-259 has a spontaneous fission 'disaster' and tends to break apart on its own in about 0.038 milliseconds.

The physicists of the time wondered if this was the limit of the periodic table. The idea that the nucleus acted like a drop of liquid was still in vogue; theoreticians had long showed that no element beyond fermium, element 100, could possibly exist. It would simply break apart before it had a chance to form.

Yet minds were changing. In 1955 John Archibald Wheeler, one of the most eminent physicists in the US, announced that there was no reason more elements couldn't be out there. At the International Conference on the Peaceful Uses of Atomic Energy, he put up a diagram highlighting the region where elements could have a half-life of more than 100 microseconds (100 millionths of a second). It stretched out to masses twice as heavy as those that had been found before. Soon, everyone was putting forward their own theories about where the periodic table would end. Richard Feynman, one of the most influential physicists of the twentieth century, used electron orbital models to suggest that the final element would be 137; other researchers suggested 172; yet more used quantum mechanics to suggest that the last true element would have 173 protons, before crashing into a 'sea' of particles with negative energy.

The only thing for certain was that any elements past fermium would have to be made by fusion, one atom at a time. It would require work at scales smaller than human comprehension, and accelerator beams more intense than anything constructed before. It was just the kind of challenge Ghiorso and Thompson relished.

Elements 101, 102 and 103 were part of Seaborg's actinides. Beyond was the realm of the superheavy elements.

CHAPTER SEVEN

Presidents and Beetles

It was midnight. The Volkswagen Beetle hurtled up the twisting roads of Blackberry Canyon, Al Ghiorso's foot anchored to the accelerator. At his side, his assistant Gregory Choppin – a 20-something a couple of years into his research career – was clutching a test tube furiously, his body battered from side to side as Ghiorso hit the apex of each bend. The Beetle was supercharged (of course it was, it was Ghiorso's), taking corners at breakneck speed as it covered the mile between the Berkeley accelerator and Stanley Thompson's lab as quickly as possible.* Below, the lights of the Bay Area drizzled into a sea of orange, but Ghiorso's attention was on the shadows ahead at the security gate. Suddenly one of the dark shapes leaped out, levelling a gun at the oncoming vehicle.

'Stop or I'll shoot!'

Ghiorso narrowed his eyes and tightened his grip on the wheel – he wasn't about to stop for anyone. Playing chicken with a loaded gun, the engineer sped on. Wisely, the security guard decided to get the hell out of the way. Ghiorso's Bug flew on up the hill and skidded to a halt outside Thompson's building, where its two passengers rushed inside.

'[The guard] was very much distressed,' Ghiorso later wrote in *The Transuranium People*. 'He came up to our lab afterwards and we apologised, but we told him we were too busy with the experiment to talk about the incident at the moment. We got away with it.'

---

*This always reminds me of San Francisco's most famous Volkswagen Beetle, Herbie, although cinema's car with a mind of its own wouldn't make its debut until *The Love Bug*, some 13 years later.

The reason for the late-night dash and the team's disregard for lab security was simple: Ghiorso and Choppin were trying to make element 101 – and they didn't have a second to lose.

Berkeley in 1955 was a strange combination of free thinking and fear. Already the Beat Generation had descended on nearby San Francisco, creating a second renaissance of modern poetry. In the coming years, rock and roll, flower power and free love would all bloom within the West Coast's premier science hub. It was a stark, colourful contrast to the monochrome of the 'loyalty review boards' that still dominated Eisenhower's America. These, spearheaded by Senator Joseph McCarthy and the House Un-American Activities Committee, thrived on destroying the careers of anyone they didn't like. In the past year, Robert Oppenheimer – the man who had led the Manhattan Project – had found his security clearance revoked for past interest in the Communist Party. It was a garbage charge and everyone knew it, but Oppenheimer had enemies who wanted him gone. Glenn Seaborg's largely neutral testimony was one of six used to show the great Oppie's 'want of character'. The moment haunted the chemist for the rest of his life. 'It was a chilling lesson,' he would write in his autobiography, 'about the consequences of making enemies, about powerful egos reacting to slights and retaliating.'

Seaborg had more luck shielding Ghiorso from the 'Red Scare' witch-hunters. Wilma Ghiorso had been a communist; she and Helen Seaborg had also been engaged in 'subversive' activities such as attending band performances on 'coloured' nights – the music was better than the stuff played for white crowds. Either fact would have easily spelled the end of the outspoken Ghiorso's career if it hadn't been for his friend's influence. 'I never detected a hint of anything that would make anyone suspect any disloyalty,' Seaborg recalled, 'yet sometimes I had to fight like hell to keep the security people from revoking his clearance.' The Beetle incident was yet another example of Ghiorso bending, breaking or completely ignoring the rules.

Three years earlier, he'd been on a cross-country flight when a brilliant thought flashed, perfectly formed, into his mind. Scrambling around, he had grabbed the back of an envelope (it was that or the sick bag) and begun writing down some calculations. Nuclear physics had, by this point, reached sizes so small conventional measurements became pointless. Instead, the researchers had developed their own strange, informal language. A 'shake', for example, was 10 nanoseconds, the time a neutron takes to cause a fission event. This had come straight from the expression 'two shakes of a lamb's tail' (a nuclear bomb, in case you're wondering, goes off in about 50 to 100 shakes). The chance of a nuclear reaction occurring wasn't measured in percentages, but in 'cross sections' – a peculiar mix of size and probability. Cross sections were measured in 'barns' (from the phrase 'can't hit the broad side of a barn'). A barn was roughly the size of a uranium nucleus: about $10^{-28}m^2$. The bigger the cross section, the greater the chance of something happening. When it came to making new and heavier elements, the reaction cross sections were getting smaller at an alarming rate.

Sat in his seat at cruising altitude, scrawling a literal back-of-an-envelope calculation, Ghiorso reckoned that if he could somehow get hold of 3 billion atoms of einsteinium and smack them with a beam of alpha particles, he would have a reaction cross section of about a millibarn – one atom of element 101 every five minutes.

Ghiorso had smelled blood. How could he resist?

* * *

There's a moment in *Iron Man 2* where Tony Stark, played by the sharp-tongued Robert Downey Jr, makes a new element based on some architectural plans his dad left him. It's the sort of thing that gets mocked regularly by scientists ... as if a movie whose villain is a dude with a magical electric whip was striving for factual accuracy.

Stark makes his beam line using anything to hand (including Captain America's shield), then twists it where it needs to go without any regard for lab safety. He makes his corrections on the fly, using little more than a wrench and scientific intuition, setting the rest of the room on fire as he shifts the beam. You don't see his private cyclotron, but I'll give him a pass – Stark is the kind of rich arsehole that probably already has one built in his garage. The crazy thing about this scene is that it's not actually that ridiculous: Ghiorso, if he'd ever owned a TV, would have found Stark a kindred spirit.

The real problem with the *Iron Man 2* scene isn't the way Stark's equipment is set up. It's not his dad's strange legacy, either (sometimes element creators have to wait until the right technology is available, even if building schematics is an odd way to preserve your ideas). The issue is that the moment Stark twists his beam onto his target it instantly turns every atom into his desired element.* Cross sections don't work like that. If they did, we'd have filled the periodic table a long time ago.

For Ghiorso and the Berkeley team, the challenges of the 'atom-at-a-time' science needed to make element 101 were staggering. First, they had to rebuild their cyclotron. Ghiorso's calculations hinged on firing a beam of 10^{14} alpha particles. The problem was that no machine in the world could *do that*. 'The assumption of beam intensity was about an order of magnitude greater than had ever been obtained,' Ghiorso later wrote, 'but I blithely assumed that this problem could be overcome.' He was right – once Seaborg obtained the funds, Ghiorso tweaked his machine and somehow managed to increase the beam's intensity 100-fold.

* The other problem, if you're feeling really geeky, is that Stark's target was made of palladium. The cross section of Stark's reaction would have been woeful.

Next, the team needed to get hold of einsteinium. 'We calculated that a one-year irradiation of plutonium would yield about one billion atoms of [einsteinium],' Choppin recalled in *Chemical & Engineering News*. Einsteinium-253 had a half-life of around three weeks. Berkeley would have to wait a full year to get the bare minimum they needed for a target and then would have, at best, a few months to shoot for element 101.

Einsteinium couldn't be placed in the machine on its own – weighing less than a billionth of a milligram, the amount was so small it could only be seen under a microscope. Instead, it was stuck to a thin layer of gold foil. Here, Ghiorso added a twist: instead of facing the beam, the einsteinium was placed facing the wrong way. Knowing that most of a beam's ions missed their target, Ghiorso was counting on the beam shooting straight through the gold without any problem. If the beam then hit the einsteinium and caused fusion, any newly created element 101 atoms would be sent recoiling from the force of the impact.

It was a brilliant idea. To discover an element, all Ghiorso needed to do was place another layer of gold foil at the back of his beam line to act as a sort of chemical catcher's net. Each night, the net could be replaced and checked for any signs of element 101 – there was no need to interfere with the einsteinium target or the beam line at all.

The final hurdle to element discovery was time. Element 101 would likely have a half-life of minutes. For previous elements, this wasn't an issue: as a rule of thumb, it takes around 10 half-lives for a reasonable quantity of material to decay away. But when you're producing a single atom at a time, a split second could be the difference between discovery and your product vanishing forever. Ghiorso knew he had to get his sample from the cyclotron on campus to Thompson's chemistry lab at the top of the hill faster than lab protocols would allow – which led to his midnight run with the Volkswagen Beetle.

The hunt for element 101 started off badly. The einsteinium wouldn't stick to the gold foil, and every time the team tried to make it adhere the whole experiment had to be restarted from scratch, recovering and repurifying every precious atom. 'We made something like five targets before we had a successful one,' Ghiorso recalled. The team tried everything they could think of – including welding it on with a blow torch – before another team member, Bernard Harvey, solved the problem with electroplating. They had a target. 'It was remarkable,' Seaborg recalled, 'in that this was the first time that such a small amount of target material was used. An invisible amount – and I mean a *really* invisible amount.'

You can see a recreation of what happened next on YouTube: a few years later, KQED, a local educational TV station, asked the team to re-enact their discovery.* Ghiorso, in a heavy lab coat, loads up the einsteinium into Berkeley's 60-inch cyclotron with forceps. Then he adds the gold catcher foil. The machine starts up and launches alpha particles at the foil. After three hours, Ghiorso and Harvey force open the thick lead safety door to the radioactive chamber. It's so heavy both men have to kick their legs against the wall just to prise it open.

A foot race begins, man against the clock. Harvey runs inside, grabs the catcher foil and sprints upstairs. Ghiorso dashes outside. Harvey hands the foil to Choppin, who drops it into a test tube filled with nitric acid and hydrochloric acid. As the foil begins to dissolve, he grabs the test tube and rushes out to Ghiorso's Beetle, leaping into the passenger side. Ghiorso hits the accelerator before Choppin has even closed the door, hurtling the car around Berkeley and racing up to the lab. The Bug skids to a halt outside the chemistry building. Quickly, the test tube is whisked from the car and

---

* Despite being broadcast in prime time, the KQED film didn't get much of an audience from the Bay Area kids: it was on at the same time as *The Lone Ranger*. The re-enactment took place during the day, while the genuine experiments took place at night to ensure there was no traffic blocking the route between labs.

handed to Thompson, who runs the tube's contents through a series of chemical reactions to get rid of the gold, acid and fission products. Then all the team have to do is load the suspected element 101 into an alpha radiation detector and wait for it to decay.

On 19 February 1955 the team got their first hit. Not wanting to sit around all night checking the machine, Ghiorso had jury-rigged the lab's fire alarm to the detector, setting it to ring if there was any sign of alpha decay. Just before dawn, as the team wolfed down their breakfast of bacon and eggs, the fire alarm sounded. Choppin recorded the incident in a 1978 high school chemistry book: 'We all gave a loud, enthusiastic cheer [...] Bernie Harvey wrote "Hooray" on the chart beside the deflection [...] When the fire alarm went off a second time, indicating that a second atom of 101 had decayed, Bernie wrote "Double Hooray", and after the next deflection, he wrote "Triple Hooray".'

Exhausted and elated, Ghiorso went home to bed, safe in the knowledge that he had discovered yet another chemical element. The next morning, he found himself summoned to the office of an irate Glenn Seaborg. There had been a fourth decay from their sample of element 101. In his excitement, Ghiorso had forgotten to unhook the detector from the alarm and had sent the laboratory's staff and students fleeing the area in panic. Ernest Lawrence had sent Seaborg a note congratulating him on the discovery... and reminding him it was lab policy not to tamper with the fire alarms.

⋆ ⋆ ⋆

The discovery of element 101 was confirmed with just 17 atoms. It was an incredible feat of science and engineering. Emboldened, Ghiorso also chose a name for it that was designed to fly in the face of McCarthyism. When the element was announced in the June 1955 edition of *Physical Review*, the team declared that it would be called 'mendelevium', in honour of Dmitri Mendeleev, the man who had come up with the periodic table. It was an audacious choice: at the height of the Cold War, Ghiorso was taking an American discovery and naming it after a Russian. 'We felt that an aggressive approach might be in order,' he reasoned, 'that if we just called it "mendelevium", maybe it would be all right.'

It soon turned out that the move would be a vital olive branch in East–West relations. A few years later, US Vice President Richard Nixon went to Moscow for negotiations with Soviet Premier Nikita Khrushchev. Seaborg was acquainted with Nixon and passed on his anecdotes about the element's discovery. After the visit, Seaborg received a package from the US embassy in Moscow. It contained a signed 1889 copy of Mendeleev's *Fundamentals of Chemistry* and a note from a Russian fan. Nixon had used the story of Ghiorso's escapades and made a powerful impression.

Mendelevium marked the end of the most successful element-hunting team in history. Shortly after its discovery, Thompson took a sabbatical to another lab and left the project. Although he would continue to play a major role at Berkeley on his return, his element-hunting days were over. Seaborg's direct involvement in the lab also began to dwindle. Already, he had been an observer at atomic bomb tests for presidents Truman and Eisenhower, had worked as a member of the Science Advisory Committee and had authored books on atoms and the wonders of the nuclear age. A few years before the mendelevium discovery, he had also taken an interest in college sports, revitalising the West Coast programme and leading to the founding of what is now the Pac-12 – the most successful college athletics conference in the US. In 1958 he would also become chancellor of the University of California, Berkeley, balancing a tricky political climate enriched by student activism and stifled by staunch conservatism.

Then, on 9 January 1961, he would reach the highest echelon science has to offer. That morning, he received a phone call from a man with a thick Boston accent. The stranger, introducing himself as Jack, asked whether Seaborg would come and work for him. After a quick family ballot (with Helen and all of the six children voting to stay in California), Seaborg decided to exercise his parental veto and accept President-Elect John F. Kennedy's invitation.

Glenn Theodore Seaborg, the Ispheming boy who had come from nothing, became chair of the Atomic Energy Commission and responsible for the US nuclear arsenal. He also continued to champion his actinide elements. In 1957 Seaborg had written to the Atomic Energy Commission urging it to create a program that could provide 'substantial weighable quantities, say milligrams, of berkelium, californium and einsteinium'. As chair, he was able to see his vision built: REDC and HFIR at Oak Ridge. The move ensured heavy element exploration would continue for at least the next 100 years.

In Washington, DC, at the heart of JFK's Camelot, Seaborg played on a stage well beyond anything he had ever encountered; only a year into his term he found himself at the heart of the Cuban Missile Crisis. In 1963 he was present as part of the negotiating team for the Limited Test Ban Treaty, which ended nuclear weapon tests above ground. During this time, he met both Khrushchev and his successor Leonid Brezhnev. When Kennedy was assassinated, Seaborg became a close adviser to the next president, Lyndon Johnson. The two men formed such a tight-knit bond that Seaborg was often invited into the Oval Office just to hang out. Under Seaborg's influence, Johnson signed the Nuclear Non-Proliferation Treaty, a worldwide agreement to curb the spread of nuclear weapons for good. Seaborg, the man who had created the element that powered the first atomic bomb, had taken giant strides to stifle the use of his most terrible creation.

Yet all of that was to come. In 1955 the Berkeley boys were more interested in celebrating their latest triumph. Gathering at Larry Blake's restaurant, they joked, laughed and took photos of themselves next to a papier-mâché effigy of the absent Thompson. Eagerly, they looked forward to discovering their next element.

But the next heavy element breakthrough wasn't destined to happen at Berkeley. It came in Sweden – and started an argument that raged for 40 years.

PART TWO

TRANSFERMIUM WARS

CHAPTER EIGHT

Nobelievium

Stockholm is a city built on an archipelago, a scattering of islets, islands and peninsulas that create a picture-postcard city of stunning harbour vistas and century-old secrets. At its heart is the old city, Gamla Stan, a crowded maze of backstreets, steep climbs and alleys just waiting to be explored. Today, the Swedish flags flutter above the Royal Palace, the sun causing the Stockholm waters to shimmer with allure. Smells of fresh cinnamon buns and chai lattes – *fika*, Swedish high tea – drift from hipster bistros and cafes. To the north, on the fringes of the city, is Stockholm University. Here, huddled away in a small corner of the campus, is the old Nobel Institute for Physics. It's a pretty unremarkable building: Brett Thornton, a geochemist at the university, told me he worked barely 100m (330ft) away for years before he realised its significance. The only clue to its importance is on a painted plaque above the door: a white C set in a blue square. Below, the words *Wallenbergsstiftelsens Cyklotron Laboratorium* glisten in sun-flashed metal. It was here in 1957 that a team of Americans, Brits and Swedes announced that they had made element 102.

Anders Källberg meets me at the entrance. An older man in a cheery blue waterproof jacket, he sets his bike against the wall and looks up at the big C. He doesn't work here any more. Nobody does. In two weeks the building will be decommissioned, its experimental halls turned into a space for art exhibitions. Källberg is the last physicist standing, here to make sure the place is safe for its new role. 'Everything is blown out,' he says apologetically. 'I've just been down to do some substance measurements in the cyclotron hall. Actually, it was kind of annoying. I detected some remnants, traces of europium-152, in drill-hole powder.' Every element has a different maximum limit considered safe. Europium's, to

Källberg's dismay, is very low. 'The amount [of radioactivity] was above the limit that would allow the public to go inside, but it's only one hundredth the natural radioactivity of uranium in concrete walls! Fortunately, the government ended up making an exception.'

We head inside the abandoned halls like grave robbers breaking into a tomb, taking an industrial elevator down to the accelerator hall. Well below ground, Källberg leads on into a large, empty room. It feels like an abandoned warehouse, a vast space waiting to be cleaned up and filled. Portable floodlights haunt the desolate lair with strange shadows. Dust has settled over the cracked laminate tiles on the floor, and the whitewashed walls are full of drill holes where Källberg has taken samples. In the far corner, scorch marks signify where the cyclotron once stood. It's an eerie place. The only sound is our footfalls, our voices and our breathing. Källberg turns and points up to the old loading cranes that hang from the ceiling, telling stories about hanging out of the shaft to press buttons and lower in equipment. Those cargoes are all gone. Only ghosts remain.

Although Källberg wasn't present in the 1950s, he knows the story of what happened all too well. As with Fermi's Via Panisperna Boys 20 years earlier, it was a small team of scientists going up against the biggest laboratory in the world. The Swedish didn't have Berkeley's vast pockets, or Seaborg's two decades of element-hunting experience. Their equipment was made by hand on a shoestring budget, their targets and beams borrowed from other labs. Yet somehow, they astonished the world – and the world struck back.

* * *

By the time of the Swedish experiment, element makers had moved beyond neutron capture. Accelerator science had improved to the point that ions of light elements, all the way up to neon (element 10), could be fired at targets with sufficient energy and intensity to achieve a fusion reaction. But firing even these small nuclei brought new challenges. As before, element makers had to shoot the beam with enough energy to overcome the Coulomb barrier – that positive repulsive force – which was protecting *both* nuclei. This meant that the energy required to get past the repulsion and form a compound nucleus was well beyond the fission barrier. Any new nucleus was supposed to just break apart instantly.

But a nucleus has another trick up its sleeve that can reduce its energy and prevent fission. Rather than undergo alpha or beta decay, it can push out 'evaporation residue' made of neutrons and photons, much like a sinking ship casting off ballast. This creates a near-instantaneous race between evaporation and fission: either the atom will cast off enough neutrons and photons to lose some 35 to 40MeVs or it will explode. Fission almost always happens first. But in the rare cases when evaporation wins, you get a new element.

As mentioned earlier, the Swedish thought they had created a new element in 1954 by bombarding uranium with oxygen ions for several hours to make the supposedly undiscovered element 100. That time, they had been pipped to their discovery by the Ivy Mike hydrogen bomb. Yet if the Swedish were disheartened, it didn't stop them from pressing on. 'We focused on trying to produce element 102,' the official lab history records. 'We received English plutonium for the study of nuclear reactions with oxygen ions, and Swiss neon-22 for uranium irradiation.' While they were certain both combinations made element 102, the low cross section defeated them: there wasn't enough material produced to prove they had succeeded with their low-budget, home-brew equipment.

Instead, they banded together with a team from Argonne in the US and the UK's Atomic Energy Research Establishment at Harwell. The collaboration's process was designed to take advantage of each lab's particular set of skills. Argonne supplied samples of curium-244 to the British; the British then painted these targets onto a thin aluminium foil and shipped them to Stockholm. Swedish engineers working under Manne Siegbahn would then bombard the foil with carbon-13 (carbon with one extra neutron). The Swedish cyclotron was similar in design to Berkeley's; the only real difference was that they used a plastic catcher's net instead of gold foil. This was a cheap and easy alternative: after an experiment, the Swedes would just take their catcher and set it on fire, melting away the plastic and leaving only the newly forged element behind.

In 1957, after bombarding 6 different targets for 30 minutes, the team claimed victory. Three of the targets showed signs of alpha decay. It was the first signs of element 102. But the chemical experiment to check their readings failed. The team faced a tricky dilemma: should they announce their discovery to the world, or keep working?

The Swedish admitted their evidence was not very strong. Part of that was down to the equipment available: precise detectors required funds the Swedish researchers just didn't have. 'The results of our irradiation were controversial,' the lab history continues, 'mainly due to the very low yield of the nuclear reactions, the broad energy spectrum for the alpha decay that was registered and the scarcely available equipment for measurement. It would have been completed if financial resources had been available.'

'These data are shaky,' Källberg agrees. 'The alpha energy spectrum [used to detect 102] was measured with a home-built 16-channel analyser. It was early in the days of nuclear physics, you couldn't just buy your equipment. The alpha spectrum was pretty crude. But, actually, I think they did a good job.'

In July 1957 the team decided to publish their results. With it came their suggestion for a name: 'Nobelium, symbol No, in recognition of Alfred Nobel's support of scientific research and after the institution where the work was done.' The choice caught the public's imagination far more than any of Berkeley's post-war discoveries. Nobel. Nobelium. Nobel Prize. It was an easy sell. Soon the name was everywhere.

Seaborg and Ghiorso were together at Berkeley when the news from Sweden came in. Neither of them believed it; the data were incomplete and uncertain. Immediately, the Americans set out to copy the Swedish experiments to check. Soon Ghiorso was convinced that Stockholm was 'completely wrong'. Nobelium, he and Seaborg joked in private, was *nobelievium*.

Researchers from both the US and USSR soon pressured the Swedish team to withdraw their claim. Checking their data, the Swedish researchers stood firm; while they admitted that Berkeley 'appeared to cast some doubts on [their] results', they refused to back down. 'We suggest that judgment on the discovery of element 102 should be reserved.'

⋆ ⋆ ⋆

Standing in the remains of the Nobel Institute, it's easy to feel the spectres of history. Instead of plaudits, the Swedish found themselves in a battle over whether they had accomplished anything at all. When, a decade later, the Berkeley team managed to replicate the Swedish findings, they didn't mention the experiment's significance.

Today, the Nobel Institute's work is all but forgotten; its claimed discovery of nobelium will probably remain unproven forever. Yet Källberg's memories aren't bitter or resentful of what came to pass. Instead, he touches the wall gently, placing an open palm on the brickwork as he remembers the people who worked here. 'It was a happy place,' he says quietly.

'Do you think they did it?' I ask.

My companion falls silent as he considers the answer. 'I think the general feeling, when I was here, was that we really produced 102 but it wasn't accepted … I get the feeling that Berkeley didn't want anyone else in the field. "The Swedes, small guys doing things with almost hobby equipment? They can't do it before us!"'

For almost 20 years, Berkeley had been the leading name in element discovery, racking up 10 elements. In that time, the rest of the world had discovered three combined. It was small wonder its team felt that they, alone, had the expertise to expand the periodic table.

They were soon proved wrong.

The Russians were coming – and they had element hunters of their own.

CHAPTER NINE

From Russia with Flerov

In April 1942 a 29-year-old lieutenant in the Soviet Union's volunteer air force stumbled on the greatest secret in the world.

Georgy Nikolayevich Flerov was fighting for his homeland at the time. In 1941 the Germans had invaded the USSR and pushed deep into the Soviet heartland. Almost a year later, the front line stretched across the country, from Leningrad (today Saint Petersburg) in the north to the Crimean peninsula in the south. Flerov was a junior engineering officer stationed in Voronezh, a short distance from the front line, who worked to repair bombers. As far as the state was concerned, he was a nobody – just one of the millions drawn from across Russia to fight in the Great Patriotic War.

Flerov's origins were humble. He was born in Rostov-on-Don to an impoverished family who couldn't afford to provide him with an education. Throughout his teens, he was a labourer, engine greaser and electrician. In 1931, aged 18, he made his way to Leningrad to work at the massive *zavod Krasny Putilovets*, a munitions and tractor factory whose strikes had swept the tsar from power 14 years earlier. Two years later, he was sent by the state to attend university – the Soviet Union needed sharp minds – and fell in with Igor Kurchatov, the Russian answer to Ernest Lawrence. Under Kurchatov's tutelage, Flerov blossomed into a promising nuclear physicist. In 1940, while investigating the different isotopes of uranium, he and another researcher discovered that fission can occur spontaneously in nature – that elements could become unstable and split apart on their own. It was an impressive discovery ... but the Axis invasion robbed him of any opportunity to follow it up.

Flerov was convinced he was of more use to his country as a physicist than as a mechanic, and had vowed to keep his scientific career on track. When off duty, his preferred way to relax was to head to the local university library and catch up on the latest research journals. And it was during one of these breaks that he noticed something was missing. Two years earlier, he had written a paper on spontaneous fission. Nobody had responded to it. In fact, nobody seemed to have published a word about atomic research *at all*. Flerov couldn't imagine that the British, Americans and Germans, with their amazing machines and vast resources, had abandoned the 'uranium problem' wholesale. That could mean only one thing: everybody else was working on an atomic bomb.

The Russians didn't have a nuclear weapons project. Instead, the state had prioritised metallurgy and heavy industry, and had dispatched its best chemists and physicists to work in industrial plants. Flerov had never been convinced this was the right course of action. In 1941 he had visited the Russian Academy of Sciences, outlining exactly how a bomb could be made, and had also written repeatedly to Kurchatov, begging the 'prodigal son' of nuclear science to unleash the power of the atom. Convinced the Americans were doing just that, Flerov knew he had to act. If the other physicists wouldn't listen, perhaps The Boss would.

Flerov wrote to Stalin.

> *Dear Iosif Vissarionovich,*
> *Ten months have elapsed since the beginning of the war, and all the time I have felt like a man trying to break through a stone wall with his head [...] Perhaps being at the front, I have lost all perspective of what science should deal with at present [...] [but] I think we are making a big mistake. The greatest follies are made with the best intentions. All of us want to do all we can to rout the Nazis, but there is no need for such hurry-scurry, no need to deal with problems that only come under the term 'pressing' military objectives [...] This is my last letter, whereupon I lay down arms and wait till the problem has been solved in Germany, Britain or the USA. The results will be so overriding it won't be necessary to determine who is to blame for the fact that this work has been neglected in our country, the Soviet Union.*

Flerov ended with a demand. A seminar with the best Russian scientists and 'an hour and a half for the report, in your presence' to plead for the creation of a Soviet atomic bomb. It was an astonishing gamble – Stalin wasn't known for taking pen pals and had a habit of 'disappearing' malcontents – but the lieutenant had been pushed too far. All he could do was wait for his reply: a commendation or a bullet.

The message reached Stalin's office in the Kremlin. Its arrival coincided with a bundle of intelligence from Lavrentiy Beria, the head of the secret police, also making the case for an atom bomb. Stalin paced his office, puffing away on his cherry-root pipe as he discussed the idea with his scientific consultant, Sergei Kaftanov. Kaftanov agreed it was 'necessary to act'. Next, Stalin called in his four greatest physicists and abused them. Why was some upstart lieutenant able to see something they had all missed? This 'Lieutenant Flerov', just a name on a piece of paper, had guts. Stalin liked guts. The USSR *needed* guts. The physicists were to start work on an atomic bomb immediately.

Georgy Flerov had baited the Great Bear into atomic action.

* * *

The first Russian atomic bomb detonated in August 1949. First Lightning, or RDS-1, was almost identical to Gadget, with the same 'Fat Man' bulbous body designed to explode inward and initiate a plutonium chain reaction. The similarity was no fluke: the Russians had stolen the plans. Rather than the Trinity test's solitary pylon, the Russian test site in the remote steppes of Kazakhstan was surrounded by wooden buildings, a fake subway station, tanks, planes and 1,500 animals to see what would happen. The animals didn't make it.

The USSR's programme had almost as many luminaries as the Manhattan Project. The lead scientist was Kurchatov. Overseeing the whole project was Beria, a man who had already ordered the deaths of thousands, perhaps millions, in Stalin's purges. Failure wasn't worth contemplating.

Flerov was another of those at the test. Shortly after his letter, he had been reassigned (much to his relief) to focus on atomic work. By the war's end he was a key part of the Soviet nuclear machine, and in mid-1945 found himself in Germany, trying to establish just how far the Nazis had come with atomic research. The answer was 'not very far'. The German bomb project had never got off the ground, in part thanks to sabotage by Norwegian chemists. The programme that did exist (described by the Allied scientific head as 'ludicrously small-scale') had already been picked clean by the British and Americans. Flerov worked to 'recruit' any German scientists that remained, now dressed as a colonel in the NKVD – Soviet state security. Unfortunately for Flerov, most of the top German nuclear scientists had already been rounded up by the British and were prisoners in Farm Hall, a manor house on the banks of the river Great Ouse in Cambridgeshire.

Among the scientific prisoners had been Otto Hahn, the man who, with Lise Meitner, had discovered fission. In 1944 he had won the Nobel Prize for it. Meitner, in one of the great moments of scientific sexism, got nothing (when the Nobel records were later opened, it was revealed she had been nominated and overlooked 48 times). An ardent anti-Nazi, Hahn had stayed in Germany but, unlike many of his fellow prisoners, refused to work on the bomb project. When word reached Farm Hall about the atomic bombing of Hiroshima and Nagasaki, he felt personally responsible and contemplated suicide. Once again, the dark side of science had taken its toll.

But Germany was a long way from Kazakhstan and yet more bombs. After First Lightning, Flerov was released from his nuclear obligations and began to turn his attention to element discovery. In 1956, after hearing about Al Ghiorso's discovery of mendelevium, Flerov used Kurchatov's cyclotron in Moscow to bombard plutonium with oxygen in an attempt to make element 102. Perhaps he even succeeded, although even Flerov admitted the results were inconclusive.

In Georgy Flerov, the Soviets had discovered their own element mastermind. But if he was going to compete with the Americans, he needed a laboratory that could rival Berkeley.

* * *

The Joint Institute for Nuclear Research (JINR) sits at the heart of Dubna, a small town two hours' drive from Moscow. To get there involves a trip down a long, single-lane carriageway that slices through heavy pine forests and cuts past a T-34 tank parked to mark the point where the Axis invasion was stopped 75 years ago. You can begin to feel the history of the place before you arrive.

Dubna is a *naukograd*, one of Russia's dedicated science hubs. Entry is past a giant metal sculpture proclaiming the town's name – a cast-steel version of the Hollywood sign – and banners immortalising its scientific heroes. The Volga River cuts the settlement in half. On the south bank, where the Volga meets the Moscow canal, a 25m (80ft)-tall statue of Lenin keeps a lonely vigil. Originally, it was accompanied by a similar bust of Stalin, but that was quietly dismantled shortly after the dictator's death.

By all accounts, Dubna hasn't changed much since JINR was established, save for a few Western touches that have crept in since the Iron Curtain fell: a McDonald's, a small supermarket, a fantasy-themed hotel. It's easy to overlook these capitalist trappings and imagine the town the first scientists must have seen when they arrived in the 1950s, drawn from across the communist world to create a centre of nuclear excellence.

The man I'm here to see was one of those arrivals. He has been here, save for lectures abroad, ever since. His name is Yuri Oganessian, and he is currently the only living person to have an element named after him.

I've seen photos of Oganessian as he was when he started work at JINR: a fresh-faced 28-year-old of Armenian descent, short in stature, with classical features, his hair slicked down and a mischievous grin always playing at the corner of his lips. As a young man, Oganessian initially wanted to be an architect, but his penchant for science brought him to the Moscow Engineering Physics Institute. Here, it soon became clear Oganessian had a gift for organising the large-scale projects that would drive post-war science. He had a creative, eager mind that could solve problems; and, more importantly, he had a talent for bringing the right people together to realise his ideas. Upon graduation, Oganessian found himself wooed by the greatest minds in the USSR. After giving his future some thought, he elected to join Flerov as his chief engineer.

It was a typically bold appointment by Flerov – the young Armenian didn't even have a PhD. The job interview was equally bizarre. Flerov sat Oganessian down and chatted to him for an hour without asking a single question about science. 'From the first meeting with him, there was a conversation,' Oganessian told the YouTube channel Periodic Videos. 'He didn't ask me about physics. He just asked me what I liked in life. Sport. If I liked the theatre, music, other things. It was just a conversation like that. Then he said "OK, OK, I'm satisfied. Thank you very much – I'll take you in my group."'

Apart from the town of their birth (Oganessian was also from Rostov-on-Don), the two men had little in common. Yet, as with Seaborg and Ghiorso, they were the perfect fit. 'This programme of superheavy elements was so fantastic for a young guy,' Oganessian recalls. 'If I had the chance to start again now, I would do it this way again.'

The feeling of respect was mutual. Flerov had in mind an audacious plan that needed someone of Oganessian's brilliance. Tired of Berkeley's domination, Flerov planned to join the race for the superheavy elements and had designed a machine – a new cyclotron – he believed could beat the Americans. Oganessian was going to build it.

The entry to JINR is at the end of a muddy road across a ruined level crossing (the warning siren always on, the barrier always open). A small checkpoint guards the entrance, where my passes are checked and authorised. A moment later, stepping through a small wooden door, I'm standing on the main boulevard of Russia's premier science facility. Some of the buildings have a fresh coat of paint and new wings as they have expanded; others have boarded-up doors and shattered windows, and seem to be in a state of general dilapidation. JINR, my guide Nikolay Aksenov explains, comprises seven laboratories, all looking at different areas of nuclear science. Funding depends on success – and some labs have been more successful than others.

The Flerov Laboratory of Nuclear Reactions is the second building on the right. It is clearly one of the more affluent laboratories, although the building itself is a blockish, whitewashed complex that takes the same form as all the others. Outside, 0.5m (2ft) metal dewars – pots containing liquid nitrogen – have been stacked ready for use. The caps to the dewars were lost a long time ago; these days, empty baked bean cans do the job.

Aksenov leads the way into the building, up the stairs to the first floor, through a secretary's office and into a long, oak-panelled room dominated by a massive conference table stacked with magazines, reports and scientific papers. At one end, rising from his desk, is the lab director, Sergey Dmitriev; standing next to him and smiling warmly is Oganessian – the man who, with Flerov, put JINR on the map. Still surprisingly spry for someone in his eighties, he hurries over to greet me in flawless English. We've never met before, but he shakes hands like I'm an old friend, promising to catch up later before heading to his office. It's hard to get over the thought that I've just met one of the most influential scientists in the world.

Dmitriev is also effusive in his welcome. He directs my gaze to a large plasma screen positioned on the wall behind me. On it is a live feed from the new cyclotron under construction, a few hundred yards down the road. It's a strangely still image: the only activity on screen is an old woman with a mop and bucket, cleaning around what looks like a giant, 6m (15ft)-wide lump of circular metal on the floor. It takes a moment to realise that these are the two dees, the electrodes that form the beating heart of a cyclotron. Once finished, it will be one of the most powerful pieces of scientific equipment in the world, joining the other five cyclotrons operated by the Flerov lab team. Currently it's still waiting on its other vital components, not least the magnet that will cause those ions to spin.

'I can't wait to see one in action,' I say. I never did see the cyclotron at Berkeley – the sticky rib sandwich laid on by Jacklyn Gates distracted me.

Dmitriev smiles. 'OK, let's go. Now is a good time to visit the main cyclotron. It works 24 hours a day and there's a queue to use it. But at the moment we can get in.'

Today U400M – the U stands for *uskoritel*, or accelerator – is being used by a private space company, bombarding their satellites with ions to simulate cosmic rays. The machine is only turned off for two weeks a year, Aksenov explains on the way, as Dmitriev leads us down a flight of stairs and along an unmarked corridor. That's because at the height of summer, the water taken from the Volga is too hot to act as a coolant. 'That's when the engineers can make repairs,' Aksenov says. 'There's another reason we do it then too: we can all take a summer vacation.'

Down the corridor, past one turn and through a small control room bedecked with monitors, readouts and flashing buttons, we arrive. The sight is like nothing else in the world. In principle, the U400M is exactly the same as the first generation of cyclotrons: an ion machine gun. It just happens to be a machine gun the size of a house that fires 6 trillion bullets per second.

When visitors saw Lawrence's 1939 cyclotron they called it a 'truly colossal machine'. It weighed 220t. U400M weighs 2,100t. At first it looks like a power plant – a large, cold concrete box with a humming machine in the centre, pipes shooting off everywhere guarded by emergency valves and metal walkways arranged to step over the crucial equipment. Yet glancing up at the huge contraption that dominates the room, you can just make out the familiar zinc battery appearance of the cyclotron's dees, sandwiched under a huge magnetic arch as if they were gripped by a clamp.

It's loud. The whole thing hums constantly as its electromagnet keeps the beam where it needs to go ... valves occasionally hiss as they release pressurised steam ... coolant rumbles from somewhere inside. White beards of frost appear at key joints where the liquid nitrogen is added from the baked bean can dewars. This coolant, along with water from the Volga, is essential. As a cyclotron demands an electrically charged projectile, atoms used in the beam have to be heated to strip them of their electrons. This means that U400M shoots an intensely hot plasma – electrically charged gas – of around 600 °C.*

'This is all pretty typical equipment,' Aksenov says, pointing around the room. 'It's just of very good construction. Pumps, pipelines, cooling water. The injection of ions is here, accelerator here, this is the beam line.' He traces his finger along the route of a pipe that weaves its way across the floor before vanishing into a solid wall. 'You focus it on a target in there. We hide the target; these blocks are to isolate the radiation, so this room is always at background radiation levels.'

Aksenov is being modest. The machine is a modern marvel: it helped discover five chemical elements.

★ ★ ★

U400M is a world away from JINR's first effort, the U300. Built to custom specifications in Leningrad and 3m (10ft) in diameter, Flerov had ordered a machine to match anything else in the world. Handing the plans to Oganessian, he had tasked his young assistant with turning his vision into a reality.

* Calcium usually turns to gas at around 1,484 °C, but the system is kept in a vacuum, which lowers the boiling point.

At first, progress was slow: none of the Russian team had any experience building a cyclotron. 'One had to be a pioneer in almost everything,' the JINR records state, 'and the only guide was one's academic knowledge and intuition. Lack of coordination and mistakes were inevitable.' Yet in Oganessian, Flerov had chosen the perfect leader. Somehow, the young Armenian kept the team together, preventing conflicts, delegating jobs and solving problems before they derailed the project. 'It was largely Oganessian's skills,' the record continues, '[that ensured] the success of its accelerator complex.' On completion, it was probably the best machine in the world for discovering new elements, capable of accelerating ions as heavy as neon.

The Berkeley element hunters also had a new toy to play with. At the suggestion of Luis Alvarez, the laboratory had collaborated with Yale University to build a heavy ion linear accelerator (HILAC), hauling its parts up the Blackberry Canyon pass on flatbed trucks. In April 1957, HILAC had

Figure 5 Transporting part of the HILAC to Lawrence Berkeley Laboratory, 1956.

begun operation. Now the Americans had the ability to shoot heavier ion beams too.

Sadly, Lawrence would never see the fruits of Berkeley's latest 'Big Science' scheme. In 1958 President Eisenhower asked him to attend nuclear treaty talks in Switzerland. Despite having a flare-up of ulcerative colitis, Lawrence agreed. He fell ill and died shortly after his return to the US. Less than a month later, the University of California decided to name their two nuclear research sites after him: the Lawrence Radiation Laboratories at Berkeley and Livermore were formed.*

With the completion of U300 and HILAC, the US and USSR teams were evenly matched. Both had cutting-edge equipment, the resources of a superpower behind them, and skilled leaders capable of element discovery.

It was the wider Cold War in microcosm – and it would prove to be just as divisive.

---

* In 1971 they became Lawrence Berkeley Laboratory and Lawrence Livermore Laboratory; in 1995 they both got 'National' in their titles.

CHAPTER TEN

The East and the West

On 3 July 1959, a little after lunch, men and women rushed out of the Berkeley HILAC building fleeing for their lives. They had seen minor lab accidents before. They had heard fire alarms go off in error or thanks to Al Ghiorso's tinkering. But this was the real deal. The entire building had been flooded with radioactive dust. Inhalation could mean death.

The lab had been unusually busy, with 27 people inside; half of the potential victims were plumbers, there to work on a tank. The first 26 evacuees to emerge were met by the Berkeley safety team, who took swabs from people's nostrils to determine their exposure. Those contaminated were ordered to strip naked. Their clothes were then tagged, placed in cement sacks and destroyed. One member of staff, Vic Viola, decided to take decontamination a step further and darted into a nearby lab to dunk his hair in its sink (Viola does not remember this, and insists he had a crew cut at the time). Clean smocks were provided to preserve dignity and stave off the chill Bay Area winds.

Finally, the last evacuee, Ghiorso, walked out of the lab. He was wearing a respirator, overalls, booties and gloves. Without a word he stripped them off, walked toward the nearest decontamination shower and let the icy water wash over him. The Lawrence Radiation Laboratory had just experienced its first major accident.

The US team had been attempting to find element 102, still convinced the Swedish claim was an error. But beyond mendelevium, the potential half-lives and cross sections of elements were tiny (the atoms were larger and more unstable), and there was only so fast you could drive a Volkswagen Beetle. To compensate for the capricious nature of his atomic hunt, Ghiorso had turned to a new method of element detection. Rather than look for the element itself, he was looking for its daughters: the known elements it would produce through alpha decay. Element 102 would decay two places back on the periodic table into fermium. When creating element 102 through fusion, all he needed to do was find fermium in his detector and the discovery would be proven.

This wasn't as straightforward as it sounds. Alpha radiation follows Isaac Newton's laws of motion: every time an atom decays and throws off an alpha particle, the nucleus also pings off in the opposite direction. If Ghiorso just used a catcher foil, as he had for mendelevium, he would never find anything: any element produced would fly off his foil the moment it decayed. Ever the inventor, Ghiorso soon had a solution. He needed to set up the element hunter's equivalent of a pool trick shot.

First, as with every other experiment, the ion beam would hit the target (in this case, curium was bombarded with carbon ions). If fusion occurred, the newly formed element 102 would be thrown forward by the beam into the end of the accelerator just like before. Here came the clever part. At the back of the chamber was a moving, negatively charged conveyor belt. Any element 102 produced would immediately be drawn to and deposited on the belt, which would then drag the element out of the target chamber and under a sheet of catcher foil. If element 102 was caught on the belt and decayed, the newly produced fermium would be flicked onto the waiting sheet of foil. Ghiorso dubbed his new toy Hades.*

The machine had been an immediate success. In 1958 Ghiorso reported that he had detected traces of fermium, which could only have come from element 102. Within a year, the Berkeley team had more data to back up their findings. It was, in Ghiorso's opinion, stronger evidence than

* Officially it was the Heavy Atom Detection Equipment Studio; unofficially, Ghiorso had chosen the name because it was, radioactively speaking, as hot as hell.

anything the Swedish had come up with – he just needed a few more experiments to be sure.

It was during those final tests that the accident had occurred. Hades was filled with helium, which served to remove all the fission products created during bombardment and get rid of unwanted radioactive noise. This required regular flushing of the system to remove any unwanted trapped gases, much like bleeding a giant radiator. During a flush, the curium target – spitting out deadly alpha particles – was left in the machine.

That afternoon, Ghiorso forgot to turn a valve and close the circuit. Instead of forcing the unwanted gases out, the helium became backed up and looked for an alternative exit. The weak spot in Hades' design was a 0.1mm sheet of nickel foil. As the pressure of the helium increased, the force against the foil was so great it punctured, rupturing the system and causing the curium to 'literally explode into a dust'. The result was like popping a balloon filled with radioactive glitter.

Hades' outer chamber hadn't been designed to deal with this kind of accident. A draught carried the curium dust up over the radioactive shielding and into the timber beams of the building's roof. There, it was picked up by the ventilation system and spread out across all 1,500m^2 (15,800ft^2) of the building in a fine mist. Within seconds, the radioactive cloud had flooded the entire complex.

Ghiorso had heard the foil being blown and turned off the valves almost immediately. Even so, a quick check with a Geiger counter showed that the curium had scattered everywhere: on the floor, on the machine, on himself. Ghiorso hit the intercom and warned Sue Hargis, the building monitor, to start a total evacuation. He then stayed, crouched at the entrance of the Hades chamber, to shut down the machine and contain the spill. Showing a cool head and immense bravery, Hargis got everyone else out and then headed for Ghiorso to hand him the respirator and overalls.

The lab was evacuated in five minutes. Ghiorso was out in 10. An hour or so later a doctor came around and recommended that everyone provide a urine sample. Only five people bothered: the researchers knew that if they had been exposed, there was nothing modern medicine could do about it anyway.

In a small miracle, no one was seriously hurt. Hargis calculated the maximum dose anyone could have received was some 1.5 Sieverts (15 million bananas) but most of the evacuees' exposure was far lower. Ghiorso, who had been closest to the dust, got lucky. His instinct to duck down meant the worst of the curium was blown clear above his head. He suffered no long-lasting effects. Viola, who received the highest dose, became a test case for health scientists, who provided him with a lunch pail in which to collect his poo for the next six months. 'The results provided the health physicists with a nice paper,' Viola recalls, 'but I was miffed because I didn't even get an acknowledgement.'

The HILAC laboratory was so full of radioactive particles that it took 30 people around three weeks to decontaminate it. Even so, Ghiorso recalled, 'for many years curium continued to be found in small quantities in obscure places in the building'. Labour, materials, clean-up costs, lost equipment and – most expensive of all – lost beam time at the HILAC meant the incident had cost Berkeley $58,500 ($500,000 today). 'It also, understandably, caused us to be quite gun-shy in the use of highly active targets,' Ghiorso added. The search for element 102 was abandoned. Instead, Berkeley began to look for the next element in sequence.

Hades was soon rebuilt and renamed (not certain it merited the name 'Heaven', the team decided it was somewhere in between and called it 'Limbo'). In 1961 it produced element 103. Ghiorso rapidly pushed for the name 'lawrencium' after the lab's recently deceased leader. Photos were taken of a delighted Ghiorso, scrawling its initials (Lw, although this was later changed to Lr) into place on the periodic table.

The Americans were confident they had discovered all of the actinides, opening the path to the discovery of the superheavy elements.

They didn't expect the Russians to get there first.

★ ★ ★

The Flerov Laboratory of Nuclear Reactions isn't known for its interior design. Just a few corridors into its beating heart and you'll pass hard concrete walls with warning lights in bent metal cages. Take a trip up its myriad stairs and you'll find yourself climbing unpaved passages that feel more like a builder's yard than a laboratory. The working ends of its machines are in rooms filled with pistons and pressure valves, sudden hisses and fluctuating needles. Signs and sirens and decontamination showers are everywhere.

Yuri Oganessian's office, in contrast, is a thing of beauty. It is a treasure trove, built up over decades, that is crammed full of memories and ideas. Most scientists' offices range from the clean and clinical to generic boltholes that lack personality. Oganessian's is in another league entirely. The room is expansive. Presidential, almost; an Oval Office of science. Upon his polished desk are papers to review, notes, pens and a calculator. There is no sign of a computer. Behind his chair are the usual family photos, including a giant A3 portrait of his grandson, the proud Oganessian boasting of his exploits in a race around Manhattan. Shelves are stacked with textbooks, prizes, awards, certificates, mementos and presents – there's even a licence plate from Roswell, New Mexico. In the corner is a blackboard, its chalk words and diagrams shielded by glass.

'This is Flerov's,' Oganessian says, gesturing to the marks. He's saved his mentor's handwritten notes for posterity. The chalk lines, reds and oranges, shoot out violently: the Russians were discussing the idea of burying a nuclear bomb underground and setting it off to produce vast amounts of neutron-rich curium. The idea isn't as crazy as it sounds. Throughout the 1960s, both the US and USSR set off nuclear bombs for a whole host of reasons – element discovery was just one of them. Oganessian's voice still quavers when he speaks of his mentor. The two men were separated by almost 20 years in age. They weren't friends, but something more – that rare bond of master and apprentice. The next room over was Flerov's, which has been converted into a small museum full of relics. My favourite is a giant stuffed crab, caught by Flerov on an expedition to Kamchatka.

Today, Oganessian oversees the lab's scientific programme. There's a joke in Russia, a belief that Armenians such as Oganessian have a creative mind that can see the world differently. It might not hold true for everyone, but it certainly holds for Oganessian. 'When you come to work for Yuri, it's not like a lab,' one of the Russian researchers told me before entering. 'It's like a theatre – and he's the director.' There's another secret I was told before my visit: 'When you meet Yuri, he'll sit you down and talk to you about what he wants to talk about, not necessarily what you *think* you want to talk about. By the end, you'll wonder why you never saw the world so clearly.'

We take our seats, thick Russian coffee in fine china at the ready. I ask him why he came into element discovery. 'I came to the group of Flerov, and the group [work] was defined by Flerov!' Oganessian laughs warm-heartedly at my naivety. 'We didn't just look at properties [of elements]. We looked at nuclear reaction, interactions, types of decay, nuclear fission, alpha emission: the wide field of nuclear physics and chemistry.'

He settles down into his chair. A grandfather telling tales. 'All this story started, for me, in 1962,' he recalls, four years after he first arrived at Dubna. By then the Russians had been trying to produce 102 for four years, coming tantalisingly close only for the data to contradict itself each time. 'I was a very young guy at the time, and the other people in Flerov's group wanted to try the experiment [for element 102] first. Then Flerov said to me: "OK, we'll try you now." I completely changed the apparatus for detection. It was my design.'

The reconfigured machine soon produced the first Russian breakthrough. In 1964 a trio of JINR scientists bombarded uranium with neon, discovering nobelium-256 by isolating its daughter, fermium – just as Ghiorso had attempted. Further experiments followed over the next few years. It was bad news for the Americans – the Berkeley team had made an error in their work, and while they had produced nobelium, they had misidentified the product. They had made two different isotopes, with two different half-lives, and assumed it was one thing.

This was enough for the Russians to denounce the American discovery and claim it themselves. Instead of nobelium, they decided the element would be called 'joliotium', after Frédéric Joliot-Curie, one-half of the dynamic wife-and-husband duo who had given the world artificial radioactivity. That Joliot-Curie had been an ardent communist and the first person awarded the Stalin Peace Prize (think a Nobel with more Cold War cynicism) was no coincidence.

The 102 debacle highlights just how crazy the politics of element discovery had become. The Swedish had (probably) made something; the Americans had tried to claim the discovery and almost killed themselves proving it; and the Russians had pointed out an error in the Americans' work before producing the first definitive proof of the element.

Element 102's discovery was the flashpoint of a period remembered today as the 'transfermium wars': a cat's cradle of disputes knitted by the US team at Berkeley and the USSR team at Dubna throughout the height of the Cold War.* The 'wars' lasted from roughly the launch of Sputnik in 1957 to the resignation of Richard Nixon in 1974. Up until element 100, perhaps element 101, there is (today) no real debate about who discovered which element.† Beyond, from the discovery of element 102 until the confirmation of element 108, *everything* is disputed.

Soon after the discovery of element 102, the Russians announced element 104. If the previous discovery had angered the Americans, the announcement of the Russians claiming

* The term 'transfermium wars' was coined by chemist Paul Karol in the early 1990s. Transfermium means 'beyond element 100'.

† There are exceptions, of course: in 1907, lutetium (element 71) was discovered, independently, by three different scientists at the same time. In the end, French scientist Georges Urbain won ('Lutetia' is the Latin name for Paris). But for almost 50 years, the Germans insisted the element was called 'cassiopeium'.

the first superheavy element infuriated them. Flerov chose to name the element after his mentor, Igor Kurchatov, who had died only a few years earlier. The move was calculated to celebrate the USSR's successes but only riled the Americans further: Kurchatov was, after all, the father of Russian nuclear science – including its first atomic bomb.

In 1967 the Soviet team claimed the discovery of element 103 from the Americans, insisting Berkeley had it wrong yet again. Instead of 'lawrencium', the JINR scientists decided to name their creation 'rutherfordium', after the discoverer of the nucleus, Ernest Rutherford. A year later, they added element 105, choosing the name 'nielsbohrium' after the Danish maven who had created the model of the atom ('bohrium', it was decided, sounded too much like the element boron, and was likely to cause confusion). The Russians had not just matched the Americans – they were in the lead.

Ghiorso's team began to play catch-up. As they reached the elements the Russians had already created, the Berkeley scientists dismissed the Dubna team's results and claimed the discoveries for themselves. Element 104, Ghiorso decided, would be the *real* 'rutherfordium'. Element 105 wouldn't be named 'nielsbohrium', but 'hahnium', after the German chemist Otto Hahn.

'The big controversy was over element 104,' recalls Matti Nurmia, a Finnish researcher who had moved to Berkeley in 1965 and had become Ghiorso's right-hand man. 'We were doing experiments on it, and suddenly the Russians beat us. First, there was chagrin within our group – the *Russians had beaten us!* We studied their work, though, and thought it was poorly documented and wasn't the quality required [for element discovery]. The first thing you measure is the half-life of an isotope. The Russians thought they saw something that lasted 0.3 seconds. We couldn't find that, but we found something that lasted 0.08 seconds – very short-lived. That was controversial – the Russians tried to defend their work and said they had made a mistake; we thought there was no scientific basis [for the discovery].'

If you're lost, don't worry; so was everyone else. By 1970 the rival periodic tables looked like this:

Element number	US name	USSR name
102	nobelium	joliotium
103	lawrencium	rutherfordium
104	rutherfordium	kurchatovium
105	hahnium	nielsbohrium

In less than 10 years of competition, Berkeley and Dubna had produced four new elements using seven different names (with one name being used for two different elements). It was chaos, with experts split on exactly which group had done what first.

The transfermium wars had left the world with two periodic tables.

★ ★ ★

You're probably wondering how this mess could have happened. Science requires papers, data and proof. Experiments must be repeated, or they become meaningless. How could two world-class labs from two different superpowers disagree so radically about something we're supposed to be able to prove?

Going over the research papers from that time isn't particularly helpful. A lot of the published material is, frankly, wrong. Many of the claims didn't have enough evidence to support them, while others had good, strong science that was dismissed out of bias. Perhaps the greatest issue was the lack of independent scientists with the expertise to understand the technical issues – with only a few labs capable of producing superheavy elements, most experts were at risk of being biased toward Berkeley or Dubna. The transfermium wars were a school-yard spat, backed by two feuding philosophies, with the history of science at stake.

For some researchers, the answer is simple. Politics. 'It was the Cold War,' explains JINR's Andrey Popeko. 'If the

Americans discovered something, our first task was to show it couldn't work at all. If we discovered something, the Americans did the same.'

Heinz Gäggeler, a Swiss chemist at the Paul Scherrer Institute who worked at both Berkeley and Dubna, agrees. 'Physicists did good experiments in Dubna, but the Americans could always criticise part of it. Of course, the Russians also had very strict opinions of the results from Berkeley that, ah, weren't always scientific.'

The politics of the periodic table stretched all the way to the White House, where Seaborg had the ear of the president, Nurmia recalls. 'The word from the administration was that they were trying to build a better relationship with the Russians and it was probably best not to aggravate them too much. There was a political intervention.' The transfermium wars had become yet another theatre of the Cold War. 'Element discovery had become political and lost its appeal as far as I was concerned.'

While the politics overshadowed the two labs' competition, they alone aren't enough to explain the transfermium wars: the Americans didn't deny that Yuri Gagarin was the first man in space, nor the Russians that the US had landed on the Moon. Beyond the Cold War drama, the transfermium wars were really over a very simple question: how do you prove you've discovered a new element?

Until the 1950s there wasn't any need to have an answer. The elements were physically present, so discoveries were usually straightforward, false claims were easily disproved and mistakes were corrected. With the superheavy elements, this was impossible: their half-lives were so finite they only existed for hours or seconds.

For Georgy Flerov, the proof required was obvious. He had discovered spontaneous fission in nature, the evidence that something had broken up without being hit by a particle. If you found evidence of spontaneous fission away from the target, it could only have come from a fusion product that had been made and, while decaying radioactively, had broken apart. Spontaneous fission was easy to detect (important for the Russians, whose detectors weren't as good as the Americans') and gave clear evidence of something happening.

The problem with spontaneous fission, however, is that it is impossible to say for certain *what* has broken up. This resulted in a string of academic papers being published, often claiming different half-lives had been found as the Russians refined their experiment. For the Americans, the papers were evidence that the Russian claims were errors – Ghiorso and Seaborg dismissed them as 'will-o'-the-wisp chases'. 'The goal had become a moving target,' they recalled in *The Transuranium People*. 'Each time a new experiment at Berkeley showed that no such activity [for a Russian claim] existed, the Dubna team would counter with a new value or some new objection to the validity of the experiments.'

The American approach was more robust. By looking for an alpha decay chain, the Americans could match their data against a known isotope's half-life at each step in the chain. 'One alpha particle,' Ghiorso declared, 'was worth 1,000 fissions.' But even so, the American technique wasn't perfect; if an alpha decay chain didn't match up with expectations, the Russians had every right to point out their experiment's flaws too.

Today, the Russians are typically painted as having less solid work than their American counterparts, mainly because modern discoveries usually hinge on being able to show alpha decay chains that link up with known isotopes. But the truth is more muddied. Gäggeler points to the work of Ivo Zvara, one of the Russian team who confirmed their discovery of element 104 through chemical analysis. 'If you go to a textbook,' Gäggeler explains, 'you can see there are very clear-cut chemical properties. You can run a chemical experiment and say, "OK, if something comes through, it *must* be 104." But the Americans wouldn't accept it. They wanted alpha decay, and that wasn't fair.'

And even with alpha decay, things are rarely clear-cut. Matti Leino, who worked at Berkeley in the 1970s, explains how complicated something that appears straightforward can actually be. 'The procedure for looking for chains is very simple. Data from the experiments come in as what we call "events". These are time-stamped: the most important parameters are time, energy and position in the detector.' From here there are four possible types of hit. You could get a real superheavy element, or a real decay; or you can get things that look like superheavy elements or decays, but which are just ghosts in the machine. 'What one looks for,' Leino continues, 'is a chain that starts with the arrival of a nucleus and continues with at least one decay, with everything happening in the same position ... in such work, one can *never* say with certainty that this decay event came from that nucleus, nor that it is a real nucleus and not some random background. One can only make a statistical estimation that the chain is real.' In the superheavy field, where everything is a long shot, this can lead to some staggering calculations. 'My rule of thumb,' Leino says, 'is that if it's a probability of less than one in a million [that the thing you witnessed happened], you should seriously start to worry.'

'The main issue was that no one had *the* decisive experiment,' Gäggeler continues. 'That *this* is the discovery of 104, *this* is the discovery of 105. Both groups made a number of experiments, all contributing to a deeper understanding, but it wasn't the final concluding experiment. Eventually, both laboratories enabled such insight that the general community said, "It's clear that elements 104 and 105 are discovered ... we use them in our own work!"' It was like the discovery of fire – nobody knows who made the first spark, but humanity has reaped the benefits ever since.

Perhaps the final word should go to a man who found himself at the centre of it all. Back to Oganessian's office and the man who can make you see the world a new way. 'All this business with superheavy elements was difficult,' Oganessian says. 'And it was at this moment we decided to investigate the "island of stability" too.'

The superheavy elements had half-lives of seconds, minutes, perhaps hours at most. While the earlier heavy elements had found uses, the newest creations vanished too quickly to harness their potential.

What if you could make them last a lifetime?

CHAPTER ELEVEN

Xanthasia and the Magic Numbers

Science is about refining things from the simple to the complex until everything makes sense. It's easy to think of atomic structure as something we fixed a long time ago, thanks to the likes of Ernest Rutherford and Niels Bohr. A nucleus at the centre, and electrons orbiting in shells: the solar system in miniature (albeit obeying different rules of physics). What could be simpler? It's an idea still taught in schools.

We've already added a layer of complexity. When Lise Meitner discovered fission, she was using the idea that the nucleus behaved like a drop of magnetic water. This is the 'liquid-drop model' and, for most things, it works rather well.

When it comes to element hunting, the liquid drop isn't enough to explain everything. Why, for example, are some isotopes more stable than others? Why do some combinations of beam and target have higher cross sections than others, making them more likely to occur? The answers lie in a final layer of complexity, first published in 1949 in two separate papers in back-to-back issues of *Physical Review*. Entirely independently of each other, the US theoretical physicist Maria Goeppert Mayer and a German group led by Hans Jensen had come up with similar, radical ideas about how the nucleus worked. Rather than compete, Goeppert Mayer and Jensen had decided to work together and publish a book on the subject. It was called *Elementary Theory of Nuclear Shell Structure*. Today, it's known as the nuclear shell model.*

---

* This is only the tip of nuclear physics: like a series of Russian dolls, the models become increasingly intricate as you get smaller and smaller. The good news is that this is as complicated as we need to go.

The idea was relatively straightforward. The electrons orbiting the nucleus were already known to form shells. Rather than just being a big blob of magnetic liquid, the protons and neutrons both had these shells too, independent of each other, related to different energy levels. This was linked to an idea called spin-orbit coupling, which Goeppert Mayer explained like a ballroom full of waltzers, all going around a room in circles (the orbit), deciding which way to step and when they'll twirl (the spin). On the dance floor, it's much easier for the waltzers if they are all circling and twirling in the same direction as it requires less energy. When she hit certain numbers – when the nuclear shells were filled – everything was at its most tightly bound. The dancers took up the ballroom but had enough space to look elegant and ordered as they glided across the floor. Add too many dancers, though, and they interrupt the flow. By spotting when there was a big difference in energy when another proton or neutron was added – where the dance was being disrupted – Goeppert Mayer could identify where the shells started and finished.

The new model of the nucleus was the biggest shift in fundamental particle science since the 1930s. The chemical world opened as if a map had been unrolled, with the number of protons and neutrons as the axes. The stable elements existed on a long, thin peninsula that arched toward the top of the chart before dipping away. On either side was a 'sea of instability' – where the whole nucleus would break apart. All the discovered isotopes existed on this peninsula; adding or taking away neutrons pushed the element toward the peninsula's edge, making the structure less stable and giving it a shorter and shorter half-life. Although in theory the elements could extend to number 173, the peninsula of stability had already run out at uranium.*

* The peninsula of stable nuclei doesn't directly correspond with filled shells, and there are even nuclei with filled proton and neutron shells way off the peninsula (such as tin-100, which has a half-life of about one second). Science is complicated.

This explained what was being seen with the superheavy elements: the Berkeley and Dubna teams were creating the elements at an imaginary cliff edge that marked the end of stability, where the peninsula fell away into the sea. The half-life of the most stable form of californium was 898 years; by fermium it was 100 days; for element 102 it was 58 minutes. Sooner or later even the most stable isotopes would be gone in less than a second. If that were the case, the superheavy elements would always be little more than laboratory curiosities. And what's the use in that?

But there was another twist. If the nuclear shell idea was true, that meant there were some combinations of protons and neutrons that would be inherently more stable; just like you sometimes need to fill the ballroom to make the dance floor fun for everyone, sometimes higher numbers of protons and neutrons resulted in a sort of nuclear balancing act.

The idea seemed almost beyond belief. One of Goeppert Mayer's colleagues, Eugene Wigner (the man who had transformed Oak Ridge into a national laboratory), admitted that the evidence for her theory was compelling, but felt the supposed moments of increased stability seemed to occur at 'magic numbers'. The name stuck. In 1963 Goeppert Mayer, Jensen and Wigner won the Nobel Prize for their work. In doing so, Goeppert Mayer became the first woman since Marie Curie to win the highest prize in physics.

The idea of magic numbers changed the face of element discovery. In both the US and USSR, the teams realised that if you could create a nucleus with a 'magic number' of protons or neutrons, the newly created element would be far more stable than anything they had made before. Continuing the theme of peninsulas and seas, they began to refer to these nuclides as being on 'the island of stability'. If, somehow, the element makers could leapfrog over the unstable elements they were currently discovering, it would change everything. 'You would have a very long half-life,' Oganessian explains. 'I can show you calculations of that time. The half-lives would not be seconds. They would be a million years. Maybe a billion years.'

The idea captured the element-hunting community. 'A remarkable shift from pessimism to optimism occurred,' Seaborg wrote. 'These new superheavy elements may be much more stable than elements 99–105 [... and] it may be possible to study their chemical properties and determine how they fit into the periodic system.'

Both Seaborg and Flerov latched onto the idea immediately. They drew up fanciful cartoons, imagining the element pioneers sailing off the end of the peninsula in small boats, exploring new territories and trying to reach the 'islands' of elements with a magic number of protons. Ghiorso, ever the contrarian, made a $100 bet with Seaborg: the elements were going to keep becoming more and more unstable and his magic island didn't exist.

No one was really sure where the 'island of stability' would begin and end, but the best theory suggested that its centre would be around the undiscovered element 114. This wasn't just a magic number: if you could create an isotope with 184

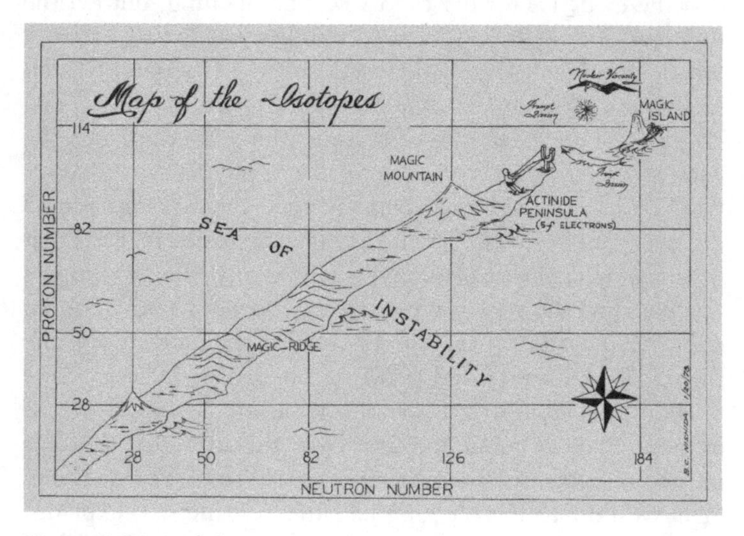

Figure 6 Map of the isotopes, showing the 'magic island' of stability, drawn for Glenn Seaborg by B.C. Nishida in 1978. The 'magic mountain' shows the increased stability of lead compared with the rest of the elements.

neutrons, it would be *doubly magic*, with exactly the right number of protons and neutrons to keep its stability. Suddenly, interest in the newly created elements blossomed again. If the element hunters could make a stable version of element 114, atom by atom, it would potentially be the greatest discovery in modern history. An amount the size of a pea could power a city.

Previously, nobody believed superheavy elements could exist on our planet any more. All of that changed with the island of stability. The solar system formed 4.6 billion years ago; if the island of stability meant half of a sample broke down every 460 million years, there would still be large enough amounts of superheavy elements created in those stellar collisions to find them on Earth.

And if that were the case, where were they?

* * *

It's possible to find trace amounts of elements beyond uranium on Earth. As early as 1943, natural plutonium-239 had been found from neutron capture in uranium pitchblende ore. We know that the Earth used to have 'natural nuclear reactors' some 1.7 billion years ago in what is now Gabon, on the west coast of Africa. Sadly, these are long gone, and the latest estimates figure you'd need 2.1×10^{32}kg of uranium ore to produce a single atom of fermium on Earth. That's 100 times greater than the mass of the Sun.

Another route heavy elements can take to end up on Earth is from the debris of supernovae and neutron stars. Through these, elements as heavy as curium have arrived on Earth through cosmic rays – basically high energy particle showers from outer space.

But neither of these finds show an element heavier than uranium is still present since the Earth actually formed.

The island of stability caused the first superheavy element gold rush. Almost overnight, the element hunt moved away from high-end ion cannons and turned to techniques almost anyone could do. 'Everyone was encouraged to participate,' German physicist Günter Herrmann recorded. 'Almost nothing was needed to perform these experiments … an intelligent choice of a natural sample and a corner in the kitchen at home could be sufficient to make an outstanding discovery.' People began to look for superheavy elements everywhere.

The different factions of the superheavy community split up, each looking in different areas for an answer. Flerov's technique was, unsurprisingly, to look for evidence of spontaneous fission having occurred in nature. If an atom of a superheavy element broke up, it would produce about 10 neutrons. If a superheavy element had a half-life of, say, a billion years, that would still mean that 1mg would produce around 400 disintegrations a second – the kind of burst that was easily detected. The only challenge was that cosmic rays (high-energy particle showers from space) could give false positives. 'We used salt mines, deep underground,' Andrey Popeko says. 'They shielded [our detectors] from cosmic rays. We set up neutron multiplicity counters.'

The Americans had the same idea. Rather than salt mines, the Berkeley team asked a favour of their local subway system, the Bay Area Rapid Transit (BART), which had just dug a 250m (820ft)-deep tunnel between Berkeley and Orinda to expand the line. In May 1970 Glenn and Helen Seaborg, accompanied by Stanley Thompson, hiked more than 2km (1.2 miles) into the BART tunnels and set up their own neutron multiplicity counters.

Some of the Russian experiments showed signs of neutron counts above background, but they weren't enough to convince the community.

Next, the two teams began to look for traces left behind by the superheavy element fission. The Russians started looking at rock samples, investigating olivine crystals – green, transparent rocks commonly found in the Earth's subsurface. Olivine is easily damaged by its surrounding environment, and if a fission fragment hits the crystal, it leaves a little track that's easily identified under a microscope. By looking at the depth of the track, you can tell how heavy an impact was – and therefore how large the atom causing it was. In America, one group started the analysis of 60-million-year-old shark's teeth, hoping for similar superheavy element traces.

Still nothing conclusive.

Flerov and Seaborg weren't willing to give up quite so easily. The elements heavier than uranium had to be there somewhere. Instead, they turned to the periodic table. As mentioned before, elements form groups with similar properties (called homologues). The undiscovered superheavy elements would, if the periodic table was right, fill up the seventh row. Element 114, if it existed, was likely to behave in a similar way to lead; perhaps it would even be *concealed* in lead. Following this piece of chemical sleuthing, Popeko went to church: specifically, to look at stained-glass windows. 'The idea was that in churches, the glass was mounted in lead frames. If there was a superheavy element spontaneously fissioning [from something hidden in the lead], then the glass would react chemically.' The Russian team also began gathering flue dust at an industrial lead processing plant for the same reason.

Using similar chemical nous, Thompson's Berkeley group started looking for element 110, which would behave in a similar way to platinum. Taking platinum samples, they managed to analyse them to an accuracy of one part per billion. Thompson also collected more than 40 large samples of natural minerals, including gold nuggets, hoping to spot element 111. Searches also began to look for stable isotopes of gases such as krypton and xenon, which were possible products of superheavy element decay.

Once again, the Russians reported signs, but the community remained unmoved.

The search turned to the bottom of the oceans and the outer reaches of the heavens. 'We looked in manganese nodules, forming in the deep sea,' Heinz Gäggeler, then working at JINR, says. 'They grow slowly, and they are made of condensed heavy metals, so Flerov started acquiring tonnes of these nodules. I got a nodule, heated it up and evaporated the lead from it.' Gäggeler didn't find anything. Down the hall, Zvara found evidence of spontaneous fission after processing 100m^3 (3,530ft^3) of brine water from the Caspian Sea, but the rate – one atom a day – was far lower than expected. Thinking along similar lines, the US team sent an expedition to California's Salton Sea, a shallow saltwater lake located directly on the San Andreas Fault, to investigate the metals bubbling up from the Earth's mantle. Again, the results were inconclusive.

Meanwhile, the Americans persuaded NASA to give them 3kg (6.5lb) of lunar rocks brought back by the Apollo astronauts. Later, they would collect data from the orbiting Skylab space station. High-altitude balloons were used to reap the skies, all just hoping for a trace – a glimmer – of a superheavy element. The Russians, not having access to the Moon, contented themselves with investigating sacks of meteorites, hoping to find some evidence of a superheavy element that had fallen to Earth. In an act of lab cooperation, Berkeley even sent Dubna 20kg (45lb) of space debris to sift through. Finally, results seemed promising: signs that could indicate past superheavies were found in the Allende meteorite, which had fallen to Earth in Mexico in 1969 and was the largest meteorite of its kind ever found on the planet. But there simply no way to be sure.

There was one near miss. In 1976 a combined team from Oak Ridge, Florida State University and the University of California, Davis announced they had found a superheavy element in nature. In certain materials, radioactivity produces a strange halo effect – a spherical zone of discolouration caused by radiation damage. Usually, slicing through these halos and looking at them under a microscope gives a neat ring structure, much like cutting through a tree, with each ring showing just how far some alpha particles managed to scatter. The new team had been looking at the halos in biotite, dark black mica, found on the beaches of Madagascar. It had halos corresponding to an alpha particle energy predicted for element 126.

If the results were correct, it would mean that tonnes of element 126 were just lying on the beaches of the Indian Ocean, waiting to be found. If that were the case, presumably the undiscovered elements in its alpha decay chain (124, 122, 120, 118, 116, etc.) were within easy reach. Better still, crustaceans in the ocean were known to regularly gobble up polonium – which meant they likely ate its homologue, element 116 too. Discovery of a superheavy element was just waiting for the first scientist to head to the Maldives, build a sand castle and enjoy a prawn cocktail. Sadly, within a year, the team's results were explained away: the '126' trace was the result of a reaction with cerium, a radioactive element found naturally in the crystal. The discovery was yet another bust.

Despite the faint glimmers and glimpses of *something*, the hunt for superheavy elements in nature had failed. 'Manganese samples, deep-sea brine samples, geological samples, meteorites ...' Gäggeler sighs. 'We did not find anything.' The teams' equipment was so good the detection limit was down to a precision of 10^{-23}g/g:* a limit beyond any other measure on Earth. Nothing was there; if there was, it would have been found.

However, there was one gem of surprising news. For over 200 years, scientists had assumed that the heaviest element left from when the Earth formed was uranium. But in 1971, while the superheavy community looked for undiscovered elements, a group from Los Alamos proved that assumption was wrong.

The team's leader was Darleane Hoffman. And she had just made one of the most remarkable – but forgotten – scientific discoveries of the twentieth century.

★ ★ ★

In the spring of 1945, 18-year-old Darleane Christian sat in front of her counsellor at Iowa State College, her arms folded and eyes narrowed. Christian hailed from West Union, a small town in the flat sea of sleepy, idyllic corn farms that make up America's heartland, where her father was the school superintendent. Petite at only 1.52m (5ft) tall, she had honey-golden hair that cascaded down to her shoulders and a rosy smile that had won the heart of several boys – men – who had gone off to war. She was the all-American girl next door. And she was not going to take sexist crap from anyone.

* Grams per gram (so you could find 10^{-23}g of something hidden per 1g sample).

Christian was in the office because she had dared to switch majors. She had initially signed up for applied art, but one of the required courses had been home economics chemistry, a subject she had never encountered before. The professor, Nellie Naylor, was an inspiration. 'I found myself more interested in chemistry than anything I had ever studied,' she later recalled in *The Transuranium People*. '[The lecturer] made it all seem so beautifully logical as well as relevant to a host of everyday problems.' Screw home economics: Darleane Christian was going to be a chemist.

Her counsellor disagreed and had summoned Christian to set her straight. 'Do you really think chemistry is a suitable profession for a *woman*?'

Christian smiled sweetly. 'I'm quite sure it is,' she replied.

Christian had graduated from her high school with the highest grades the institution had ever recorded. In her spare time, she had breezed through advanced mathematics and trigonometry correspondence courses, taken up the saxophone and played basketball for her school (which, given her height disadvantage, was a feat unto itself). Her heroine was Marie Curie, a scientist who had discovered elements, explored radioactivity and won two Nobel Prizes all while raising two children. Darleane Christian was a real-life Lisa Simpson, with the cartoon character's same stubborn streak and tenacity. If Curie could do it, why couldn't she?

The counsellor was prepared for Christian's attitude and played a trump card: even if Christian made it as a chemist, there was no way she'd find a job in the chemical industry. She'd end up a chemistry teacher – and female teachers were expected to resign if they married. Christian was far too interested in boys to end up a spinster. The counsellor expected her to give up and stick to art.

Christian didn't flinch. 'So,' she decided with a defiant grin, 'I'll just never teach. I'm going to follow Marie Curie's model. I'll marry if I want and I'll have children if I choose.'

The counsellor had run out of objections. That was fine for Christian: she wasn't asking permission anyway. For the rest of her undergraduate course she was usually the only woman

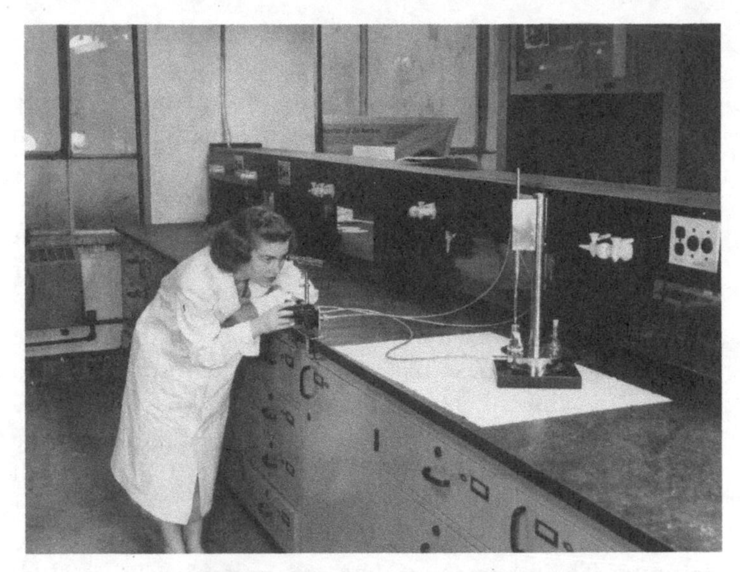

Figure 7 Darleane Christian at the Ames Laboratory, 1950.

in the chemistry lectures, even though the Second World War had plucked a generation of fighting men from their classes. She excelled, of course.

Short on money, Christian paid her way through college working summers as a waitress and then as a bank teller. By 1947 she was sick of doing boring summer jobs and applied for a research position at 'Little Ankeny', the Ames Laboratory at Iowa State. This was a clutch of small, single-storey buildings on the edge of campus, and college legends spoke about secret experiments and strange, late-night flashes of light sparking from inside. The stories were true: Ames had been another of the Manhattan Project's satellites and was focused on improving uranium production. Christian applied and, like Al Ghiorso before her, got a job building Geiger counters. Unlike Ghiorso, Christian loved it – she was being paid $170 a month for something she'd have happily done for nothing.

At Ames, barriers were thrown in her career path – barriers that seemed conveniently forgotten for men. Security passes, for example, required a person to have three initials. Christian had no middle name. But if the lab workers thought that was going to deter her, they didn't know who they were dealing with. With a shrug, she put 'DXC' on the form and told them she now wanted to be called Darleane Xanthasia. Nobody was dumb enough to argue.

Christian was soon heading to be a nuclear chemist. But the roadblocks only became harder. Two years into her course, her father died. Despite her grief, she plucked up the composure to head to the university and ask if she could be excused from the next day's quantum chemistry exam to arrange the funeral. Her professor – in an act of cruelty masked as compassion – forced her to take the test on the spot. Tears in her eyes and mind elsewhere, Christian still got a B.

Without her father's income, Christian's family were destitute. Their home was lost and possessions auctioned. Christian was the only breadwinner in the family and arranged for her mother and younger brother to live with her on campus. After graduation she started to complete her doctorate but found that her cramped quarters weren't ideal for study. Instead, she spent time each evening with Marvin Hoffman, a man who had exactly what a woman like Darleane was looking for: late-night access to a synchrotron for photonuclear-induced Szilard–Chalmers reactions. In December 1951 she finished her PhD and married him.

The roadblocks continued. Now Darleane Hoffman, the young chemist worked for a time at Oak Ridge, before moving to Los Alamos to lead a nuclear chemistry group. On arrival, the staff at the personnel office looked at her with barely hidden contempt. 'There must be some misunderstanding,' she was told. 'We don't hire women in that division.' It took a month of waiting around trying to sort out the paperwork before she ran into her supposed supervisor at a cocktail party. The supervisor immediately contacted the personnel office and got things straightened out. Yet still the roadblocks continued. Now at least acknowledged as being supposed to be at Los Alamos, Hoffman's clearance was mysteriously lost.

After another three months on the sidelines, Hoffman grew tired of waiting and called in the FBI. The 'lost' credentials turned up almost immediately.

While Hoffman had been trapped in bureaucratic hell, she missed out on the discovery of two elements. Los Alamos had been the first stop in the US for the filters collected from the Ivy Mike hydrogen bomb test (see Chapter 6). By the time Hoffman was cleared, the filters had been analysed and passed on to other labs – which had resulted in the discovery of elements 99 and 100. It was, for Hoffman, an unforgivable slight that had cost her a place in history. 'I missed being a discoverer of einsteinium and fermium … while I was sitting in a small apartment in Los Alamos raging against the system,' she later wrote in in *The Transuranium People*. 'I will never again trust personnel offices, not just for saying "we don't hire women in that division", which was untrue, but for their general insensitivity, incompetence and bias.'

It would be a lie to say it got easier. Jacklyn Gates remembers Hoffman telling her of times she walked into machine shops where *Playboy* pin-ups had been left on walls deliberately to intimidate her; of casual sexism from people who assumed she was a secretary; of times she was underestimated because she was just a 'sweet little lady'. Anyone who thought Darleane Hoffman would be a pushover or would back down soon thought twice. Thanks to her, any notions the Los Alamos personnel office had about women in science were quickly dispelled.

Hoffman's abilities matched her iron will, silencing critics with some of the most brilliant chemistry of the twentieth century. She became an expert in fission, in new isotopes and in how bacteria interact with metals. She would also become a tireless activist for women in science (although her greatest pleasure was winning awards as a *chemist*, not just because she was a woman). Ask any of the latest generation of US nuclear chemists working today, male or female, who helped them the most and Darleane Hoffman's name will come up time and time again. For a decade, despite being 13 years his junior, Hoffman's picture was hung on Glenn Seaborg's office wall as a source of inspiration (eventually he replaced it, no doubt to Helen Seaborg's annoyance, with a picture of him meeting the movie star Ann-Margret).

By 1971 Hoffman had 20 years of chemistry experience behind her, and was a brilliant researcher who wore distinctive cat-eye glasses as she oversaw some of the most painstaking research in nuclear chemistry. By this time, the superheavy element searches were raging across the world. Hoffman set her sights far lower: she just wanted to prove that there was something from when the Earth formed that was heavier than uranium. For this, the obvious target was plutonium-244 – the isotope first detected by Los Alamos after the Ivy Mike tests, with a half-life of 80 million years.

Hoffman's plan of attack was simple. She would obtain a sample of an ore that had an unusually high concentration of heavy elements and test it. Searching across America, her team found a pre-Cambrian bastnäsite, a 4.6 million-year-old lump of rock, being processed by the Molybdenum Corporation of America to make magnets, lasers and parts for self-cleaning ovens. The rock was being mined for cerium oxide, as it had 500,000 times the normal quantities of CeO_2 than normally found on Earth.

A diligent chemist, Hoffman knew that if plutonium was going to be anywhere, the bastnäsite was it. Better still, the way the CeO_2 was extracted wouldn't remove any of the precious plutonium. She called up the Molybdenum Corporation and asked if she could have all their waste material. The company were happy to oblige. Hoffman partnered with another researcher, Francine Lawrence, and began the delicate task of processing the sample. This was painstaking, careful chemistry – as tricky as anything Stanley Thompson had pulled off during the Manhattan Project.

Once finished, Hoffman shipped the sample off to General Electric in Schenectady, New York, where a friend had one of the most sensitive mass spectrometers (a piece of kit that weighs the different components of a sample) in the world. If plutonium was there, they would find it.

That night, Hoffman went to the open-air opera in Santa Fe. She cast her eye to the cosmos, up to the glittering sparkle of alien suns, each furiously creating elements and ready to scatter them across the worlds. 'As I looked out at the bright stars in the clear New Mexico sky behind the stage,' she later wrote in *The Transuranium People*, 'I somehow had the feeling that this time we would find remnants of the elusive Pu-244 remaining from the last nucleosynethesis of heavy elements in our galaxy, some five billion years ago.'

She was right. When she returned to the lab the next day, the results sent back from New York were unmistakable. The sample contained 8 femtograms of plutonium – a sample that would be invisible under even the best microscopes of the time. It was a dot of pure metal that dated from the very origins of our planet, arriving from a supernova explosion somewhere near our solar system and caught in the moment a cloud of gases coalesced and cooled to create our home.

The small-town Iowa girl, through skill and smarts, had achieved something her idol Marie Curie would have been proud to claim as her own. She had found the heaviest rock on Earth – and with it, she had touched the origins of us all.*

Perhaps unsurprisingly, after her success Hoffman found herself surrounded by scientists asking why she didn't follow up her success of detecting natural plutonium for the first time by looking for superheavy elements in nature. Her reply was easy: finding plutonium was difficult enough. Trying to find an element whose atomic number, weight and chemistry could only be guessed at? That was nearly impossible.

---

* It wouldn't be heavy element science if this weren't disputed. Later research has been unable to reproduce Hoffman's achievement, and there are questions over whether her experiment worked at all. There's also some debate (it's impossible to be sure) as to whether the plutonium she found was the result of cosmic rays, or really dates to the formation of Earth. It may not be very scientific, but after all she went through, I'm going to give Hoffman the benefit of the doubt.

CHAPTER TWELVE

Life at the Edge of Science

Lab life has rhythms, riding its own beats and twisting in its own revolutions as people come and go. By the 1970s Berkeley Lab had escaped the shadow of anti-Communist sentiment and evolved into an eclectic mix of experiments that had won eight Nobel Prizes. Under Lawrence and McMillan (who finally retired as director in 1972), the house on the hill had become the undisputed king of experimental physics, computing, energy and even cutting-edge biology. HILAC was gone, metamorphosed into Super-HILAC (the team would have preferred something even more powerful, but the Vietnam War meant funds went elsewhere). Now, Al Ghiorso was running the heavy element show, the lab inflected by his own, unique brand of crazy brilliance. In place of Seaborg's detailed notes, the lab's logbooks were home to the group leader's doodles – abstract, vivid, swirling kaleidoscopes of colours reminiscent of Wassily Kandinsky or Henri Matisse. Ghiorso's voice was regularly heard down the hall, yelling out liberal gripes in protest at how the military had sucked up all the funding for his research.

Below the Berkeley Hills was much the same: a hive of excitement that was alive with a healthy anti-establishment vibe still lingering from the Summer of Love. Music from David Bowie and John Lennon to Roberta Flack and Stevie Wonder filled the streets; and the nearby Oakland Athletics baseball team won the World Series three times. Across the Bay, you could see the old federal prison on Alcatraz, covered in graffiti after a 19-month occupation of the island by Native Americans in protest at the treatment of certain tribes. Beyond, the Golden Gate Bridge was a world-famous landmark. Things around San Francisco were looking up – and being a scientist was finally cool.

The May 1973 issue of *Ebony* illustrated how science and style mixed. There were glossy photos full of screaming flares and excessive collars, copious adverts for cigarettes and wonderful, lavish typefaces. The cover star of the month was jazz singer Nancy Wilson. The main feature was on Sammy Davis Jr. singing at the White House. But its profile was a man its staff writer described as 'relatively obscure, unpretentious yet supremely self-confident [...] a sort of hip, scientific [individual].' His name was James Harris. And he was the first African American to discover an element.*

In 1955 Harris had been a 23-year-old army veteran searching for a job. He was born in Waco, Texas, raised by his mother in Oakland, California (his parents divorced when he was young) and had a degree in chemistry from Huston-Tillotson College in Austin, Texas. Harris knew it would be tough walking back into civilian life but hadn't appreciated just how ingrained racism in science would be, even in the liberal haven of the San Francisco Bay. He was turned down a dozen times by interviewers shocked when he walked in, or by secretaries insistent he was applying to be the janitor, not a skilled chemist. Once, he was given an aptitude test so simple a child could pass it – basic addition and subtraction. Harris had looked at the sheet, passed it back to the secretary and, firmly but politely, told her he didn't need a job *that* badly.

Eventually he found work as a radiochemist for a company in the Bay Area, before moving to Berkeley Lab five years later and joining Ghiorso's team. Harris was an oddity – like Ghiorso, he never had time to get more than a bachelor's degree – but he was the man chosen to clean the team's targets, a process that took 22 arduous chemical separations playing with a mere 60μg of radioactive metal. It was the

* He was not the last. In 2009 Clarice Phelps aided in the purification of berkelium, which led to the discovery of element 117 and confirmation of element 115. You can read more about that in Chapter 20.

most delicate part of the Berkeley operation. That meant Harris had to be one of the best chemists in the laboratory and, by extension, one of the best in the world.

Another face found in the Berkeley Lab was Glenn Seaborg. The elder statesman of elements had been away in Washington for 15 years and had become of the most eminent scientists in the world – his biography was the longest entry in *Who's Who*. During his Washington tenure, he had also completed his transformation into a consummate politician. During one of his final government hearings, a Louisiana senator had tried to corner Seaborg with a *coup de grâce*: 'Dr Seaborg, what do you know about *plutonium*?' A younger Seaborg might have retorted that he'd discovered it; older and wiser, he merely smiled and promised the senator he knew a little.

Seaborg was content to let Ghiorso run the element hunt. Instead, he fell into a comfortable routine. Each morning, he would walk up and down his infamous steps. Next, he would return to his office, keeping the door open for any student who popped by. If one did, Seaborg would immediately drop whatever he was working on – often replies to the president of the United States – to help answer their question. If a student suggested a wild, fantastic idea that Seaborg knew wouldn't work, he still told them to try it; the experiment would fail, but the student would get a chance to understand why. Once his charges were set for the day, Seaborg would then tour the lab, his head poking through the hallway doors and asking in a gruff Midwest voice: 'So, what's new?' Berkeley Lab's staff ensured there was always *something* to tell him.

Often the news came from Matti Nurmia. By now, the Finnish researcher was firmly entrenched as Ghiorso's closest associate. In 1968 Nurmia had also brought over two students from his group at the University of Helsinki, the wife-and-husband duo Pirkko and Kari Eskola, who took over the painstaking work of analysing the endless stream of data from the laboratory computers. The Finnish outnumbered the Americans three to two. Matti Leino, another Finn who later joined the Berkeley team, joked that it took a Nordic scientist (and later, Japanese scientists) to keep pace with Ghiorso.

As with Darleane Hoffman, Pirkko Eskola found the science culture of the US tainted with sexism. 'Women scientists were not that common, at least not in the US,' Nurmia recalls. 'Mrs Eskola was a lovely blonde lady. She'd encounter all kinds of things. She'd call up another lab about scientific matters, and people would ask "Are you a secretary?" A woman in nuclear science was a rare thing.' Eskola was more than a match for them. She had no trouble standing up to Ghiorso either; while he constantly wanted to try something new, Eskola usually took the scientific high road and wanted more data. The result was, according to Nurmia, 'rapid-fire discussions' between the two, often with Eskola walking away the victor.

The element hunters knew how to have a good time. When an element was created and confirmed to Ghiorso's satisfaction, the team celebrated with a 'HILAC punch party' – excessive drinking, joke presentations and a

Figure 8 The Lawrence Radiation Laboratory, Berkeley team, April 1969. From left to right: Matti Nurmia, James Harris, Kari Eskola, Pirkko Eskola, Al Ghiorso.

wall-sized game of snakes and ladders. Rumours of crazier exploits still echo in the Berkeley halls to this day; one recounts that Ghiorso used to stuff radioactive material into tennis balls (the rubber was just thick enough to shield the radiation) and bat them between colleagues.

Despite the fun, the search for superheavy elements had run dry. The island of stability and the elements surrounding it seemed like ghosts, and repeated attempts to make them had, like the hunt in nature, failed. The only claims were coming from an Israeli–British team at CERN headed by Amnon Marinov, who were churning out a seemingly endless ream of papers claiming they had discovered element 112. To quote one superheavy researcher: 'Everyone knew it was bullshit.'*

The problem was neutrons. As mentioned before, the element hunters were using a technique where the nucleus discarded neutrons to stave off fission. This meant that any isotope created would, inevitably, have a relatively low number of neutrons remaining. When looking for the island of stability, this was a disaster. The first viable magic number of neutrons was 184. Even using the best beam and target available, the closest the element makers were likely to get was 173 neutrons: 11 shy of the island. The reactions in the lab were showing a 'drift to the north' on the chart of nuclides: instead of approaching the island, they were just making elements too unstable to detect.

The imagined boats navigating the sea of instability had broken rudders – and Ghiorso had run out of ideas.

★ ★ ★

Dubna's JINR had its own scientific rhythms. The teething years had passed, and the institute was proudly claiming

* Marinov later claimed to have discovered evidence of element 122 ('or a nearby element') in nature from a sample of thorium – a feat of detecting one atom in a trillion. Once again, the superheavy community dismissed his paper.

scientific victories that won Lenin medals and Nobel Prizes. Staff at Dubna had explained Cherenkov radiation (that blue glow seen in atomic reactors), explored new areas of quantum physics and pioneered Russian computer science. Things were going well.

Even so, life in Russia was vastly different to California, and visitors from the West usually experienced a culture shock on arrival. Scientists would find themselves flanked by stone-faced minders as they walked around town, and tales abounded of how Dubna's only hotel had an entire floor given over to the KGB, whose spies listened in on the rooms each night. Yet some visitors have also spoken of an incredible, unbreakable bond of companionship away from the prying eyes of the security services. On one research trip, a gaggle of visiting Americans went camping with their Soviet colleagues. Once the group were alone in the woods, the Russians produced a transistor radio and tuned into an illicit frequency. Soon, the entire campsite – Russians and Americans alike – were twisting and jiving to Johnny Rivers' 'Secret Agent Man'. 'Once you get the governments out the way and let scientists speak,' one American chemist told me, 'you find that you may have different cultural languages, but the same technical language.' Johnny Rivers transcends all borders.

Georgy Flerov was still in charge of his laboratory, steering it in his own, inimitable fashion. Heinz Gäggeler laughs as he recalls his first encounters with the Russian element tsar in 1975. 'He liked to talk,' Gäggeler says. 'He was quite often walking up and down in front of his office. If he saw someone, he asked what they were doing.' Once, after the Swiss researcher briefed him on his project, Flerov asked Gäggeler about his hobbies. 'I said I was interested in alpinism. Flerov liked mountaineering too. So, I told him I was interested in climbing Lenin Peak [at 7,134m (23,000ft), one of the highest mountains in the Soviet Union]. At that time, you needed the help of the Ministry of Sport in Moscow to go to such an exotic place – I would have had no chance as a foreigner.' Gäggeler's passes arrived soon after – and he led a Swiss team to climb the mountain later that year (bad weather prevented them from reaching the summit). 'Georgy opened the door to the ministry for me. I was a nobody, but he was very famous.'

The person closest to Flerov was Oganessian. By the mid-1970s the duo had worked together for 15 years. Although they weren't exactly friends, they had formed such a close working relationship that they were, at times, inseparable. 'He opened me to science, to physics,' Oganessian told the YouTube channel Periodic Videos. 'At 6 p.m. I would come home. At 9 p.m. I would get a phone call [from Flerov]: "What are you doing?" I would say I was doing nothing. "Come to me, please." Every day. And every day, from 9 p.m. to 10 p.m., we'd have an hour's discussion. Sometimes he called me in the early morning, and he'd just say: "I'm very sorry to call you so early …"' It was Oganessian's wake-up call – if Flerov was working, so was he.

Despite his friendliness, Flerov – typically referred to just by his initials, GN – was a stickler for rules. He couldn't abide independent research and ordered his staff not to 'dabble in zoology'. Anyone who deviated from his instructions was branded a 'guerrilla'. 'If guerrillas were found out or, even worse, proved to be successful,' the Dubna history warns, 'GN acknowledged the importance of their work in a cool and indifferent manner, without a hint of encouragement or praise.'

At meetings, Flerov kept a gong; once it had been sounded, the topic was settled and it was onto the next item. 'He was a single-minded, spirited and straightforward man,' JINR's history records, 'a man who would always rush to the charge rather than try outflanking manoeuvres. A man who did not take kindly to meandering or deviating from the task at hand, a man who jealously guarded his flock from straying.'

The Russians had reached the same stumbling block as the Berkeley team. But, unlike the Americans, one researcher did have an idea about how to rekindle the hunt for superheavy elements. Unfortunately, Flerov refused to accept it. *Clang.* Onto the next topic.

The idea Flerov had discarded was called cold fusion. But Oganessian thought his mentor had made a mistake. As

Flerov had done with his letter to Stalin, Oganessian decided his only option was to gamble his career on being right.

The young Armenian turned guerrilla. 'Cold fusion, my lovely reaction,' Oganessian remembers. He smiles at its very name. 'My cold fusion. This … this was something really new.'

★ ★ ★

Cold fusion is a term likely to make most scientists roll their eyes. It's a name that's also given to a nuclear reaction that would occur at room temperature – pure science fiction that gripped the world in the late 1980s. In element discovery, however, cold fusion is very real.

The idea first emerged in the mid-1960s from Ya. Maly, a Czech scientist working at Dubna. The concept was simple enough. So far, elements had been created by taking a light element and shooting it at something heavier. But now the technology was available to fire heavier projectiles. Why not use elements closer together on the periodic table?

This wasn't thought to be as simple as it sounds. The Coulomb barrier gets stronger when two similar-sized nuclei are pushed against each other – much like it's easier to push a small magnet against a large one than two magnets of roughly the same strength. This, everyone thought, meant you needed to fire your projectile at a higher energy to force it through – and higher energy meant a greater chance of fission.

But what if that isn't what happens at all? Think about two drops of water splashing into each other. As they come together, there is a moment when the new droplet is forced to change its shape to adapt. The same happens to two nuclei. 'The microscopic correction to the liquid-drop mass,' wrote US physicist Ken Moody in *The Chemistry of Superheavy Elements*, 'acts as a heat sink, stealing excitation energy away from the compound [newly formed] nucleus.' This means that the energy required to combine two similar-sized nuclei is actually two to three times *less* than for lighter ion reactions. Less energy required means less need to evaporate neutrons for the nucleus to become stable. More neutrons mean more stable elements.

Cold fusion has its flaws. If light-ion-induced reactions are a bombastic slamming together of two nuclei, a SWAT team kicking down the atomic door, cold fusion is a surgical ninja strike, stealthily squeezing *just* under the Coulomb barrier with the minimum energy possible. To succeed, the two nuclei must strike each other perfectly: otherwise, they just end up bouncing off each other. Cold fusion also requires a specific choice of target to make the whole trick work. In practice, this means you can only use targets made of lead or bismuth (lead, especially Pb-208, is doubly magic and thus extra stable; Bi-209 is, obviously, very close and gets some of the same benefits).

Nobody really believed it would work, but in 1973 Oganessian was willing to take the risk. When Flerov went on holiday to Siberia, his right-hand man assembled the team and got to work. 'He was on a hiking vacation,' Oganessian remembers, laughing at his mischief. 'He didn't believe the idea. But, finding myself in a situation where he was out, I started to do the experiment.'

Oganessian decided his proof of concept would be to make fermium-244, an unstable isotope that has a half-life of around four milliseconds before it completely self-destructs in spontaneous fission. 'Normally, fermium is produced by neutron capture, through uranium,' Oganessian continues. 'We wanted to fire argon into lead. It was thought to be impossible – argon was supposed to be too heavy for fusion. But I made a set-up that allowed me to adjust the intensity of the ion beam.'

Argon pummelled a lead target for five days. 'The result was amazing and stunning,' the JINR lab records report. 'The detectors were riddled with fission fragments.' As predicted, the fermium had undergone spontaneous fission. But the half-life Oganessian detected wasn't four milliseconds. It was 1.1 seconds – more than 250 times longer. The team had created an entirely different isotope.

'It was so big!' Oganessian exclaims. 'It was seconds! I was very surprised and excited. Even at the intensity I had to use for the beam, the cross section was a thousand times greater … at that moment, I knew we had cold fusion!'

On Flerov's return to the lab, he gave Oganessian the same treatment as any other guerrilla. 'Not only did he not show that he was particularly glad,' JINR's records state, 'but he actually looked indifferent.' Flerov ordered the lab to go back to its previous programme. *Don't dabble in zoology.*

A short time later, the president of the USSR's Academy of Sciences visited the laboratory. Flerov summoned Oganessian up to his office and motioned to him. 'He produces [elements beyond uranium] in their tens of thousands,' he told his visitor, almost offhand. The president realised what it meant and grabbed the physicist by the shoulders. 'He gave the lucky beggar three kisses on his cheeks,' the JINR history records. 'That was his reward for insubordination.'

★ ★ ★

With cold fusion, a whole new realm of possibility had opened. In 1974 Oganessian used the technique to shoot chromium ions into lead. The result was spontaneous fission – and the first signs of element 106. Excitedly, the Russians prepared to announce the element at an upcoming conference in Nashville. Although they didn't say why, they made it known to the world that Georgy Flerov himself would attend.

Almost simultaneously, the Berkeley team were also preparing to announce the discovery of element 106. By now, the team's composition had changed. Ghiorso was still in charge, with Nurmia at this side, but Harris and the Eskolas had departed. In their place were German-born physicist Mike Nitschke and another married couple, this time from Yale University: Carol and Jose Alonso. Joining them in the chase were a group from Lawrence Livermore National Laboratory led by Ken Hulet and Ron Lougheed. Hulet had been a health chemist at Berkeley and had been drawn into the superheavy world as one of Glenn Seaborg's post-war apprentices.

Livermore was no longer just an offshoot of Berkeley; it was one of the US Department of Energy labs where the US designed its nuclear weapons. Hulet's interest in superheavy elements wasn't a full-time pursuit – it was a side hobby, something the government was happy to support if it meant they kept the brightest minds working for them.

The US team's attempt to find element 106 had started off well. Hulet and Lougheed prepared a californium target, while Ghiorso ran checks on his oxygen-18 beam. Meanwhile, Jose Alonso tested the team's latest computer by asking it to run over some data from 1971. To his surprise, the computer suggested that the Americans had made element 106 already and failed to notice. This time, when the machine produced an alpha chain that chimed perfectly with known isotopes, the Berkeley–Livermore team spotted it immediately.

Almost simultaneously, the Russians and Americans had discovered the same element – and both wanted to announce it first. The Nashville conference attendees could sense something was happening. Rumours of a new element started to swirl. Tennessee had become the set for a Cold War thriller.

The only member of the Berkeley team in Nashville was Carol Alonso – the rest had stayed home to get more data. On the second day of the conference, she and the other speakers were invited to take a cruise down the Cumberland River on a paddle wheel boat – a majestic floating palace straight from the pages of Mark Twain. Alonso, the only woman on the boat, soon found herself besieged by researchers eager to know if the rumours of element 106 were true. She confirmed them, and, taking position next to the giant wheel at the back of the boat, turned spymaster. From her hiding place, she sent four friends to subtly ask Flerov if the Russians had also made element 106. Flerov was wise to them. 'No,' he'd teased one researcher. '[We're announcing] 108!'

That night, back on dry land, Alonso phoned Ghiorso for orders. The maverick decided to hold off announcing the US discovery, telling her 'it would be better to let the Russians go out on a limb and just watch to see if it got chopped off'. The next day, Alonso – still playing superspy – managed to sneak an advanced copy of Flerov's paper from the conference organiser

Figure 9 The Russian and US teams meeting in Dubna, USSR, 1975. From left to right: Yuri Oganessian, Georgy Flerov, V. A. Druin, Al Ghiorso, Glenn Seaborg, Ken Hulet.

and confirm his plans. When Flerov's speech was delivered and the Russian discovery of element 106 was revealed, Alonso was able to play it cool and ignore the claim entirely. It was a dirty trick – depriving the Russians of the oxygen of the publicity and credibility from the US's own results – but it worked.

A week later, the Russians visited Berkeley. There, both teams told each other of their element 106 experiments in full. The Russians were impressed by the thoroughness of the US team; the American team were less impressed by the Russian effort, but had no obvious grounds to object to the validity of their claim. For the first time, the discovery of an element had ended in a stalemate.

The teams had been competing for 15 years. Neither side – Seaborg and Ghiorso, Flerov and Oganessian – had any interest in continuing to fight. Both Berkeley and Dubna agreed that nobody would suggest a name until the results were confirmed.

Element 106 had been discovered, but its space on the periodic table would remain blank. The hot phase of the transfermium wars was over.

CHAPTER THIRTEEN

The Atoms That Came in from the Cold

Two cars rushed through the San Francisco countryside at breakneck speed, wheels almost touching as they jockeyed for position on the road. At the wheel, Al Ghiorso battled to keep his supercharged Volkswagen straight, his team clutching onto their seats as he revved the engine past its supposed limits. Next to him, packed like sardines into their own vehicle, Flerov, Oganessian and impassive KGB agents raced past, barging out in front and blocking the Americans' path.

Ghiorso focused and twisted the wheel, the smell of burning rubber permeating the air as he bobbed and weaved, trying to squeeze his way around the Soviets. Finally, teeth gritted in sheer determination, he pulled the Beetle out onto the shoulder and floored the accelerator. Rubble, smoke, dirt and dust plumed out from behind the car as the Americans, slowly but surely, overhauled their Russian counterparts. Ghiorso cast a glance over at his defeated rivals, smiling in triumph before turning to look ahead. There was nothing but a solid brick wall. Ghiorso hit the brakes, but they didn't work. The Berkeley boys were hurtling to certain doom, out of control …

Ghiorso woke up, the nightmare fresh in his mind, out of breath from his imagined panic. He was safe in his home in Berkeley, Wilma at his side. It was 17 July 1976. Climbing out of bed, he went and had a shower. *Out of control.* He closed his eyes and groaned. His subconscious had just shot down the greatest discovery of his career.

For the past 18 hours, Ghiorso had been convinced that his team had found the island of stability. Partnering with Livermore, the Berkeley team had been bombarding curium (19 years since the HILAC explosion, Ghiorso was happy to use the radioactive material again) with a new beam – calcium-48. This was a brilliant idea. Natural calcium is a bad isotope for element hunting – it doesn't have enough neutrons. But a quirk of nature means that around 0.19 per cent of natural calcium is Ca-48, with eight extra neutrons – plenty to evaporate and try to stabilise the newly formed element. Even better, Ca-48 is 'doubly magic': both 20 (protons) and 28 (neutrons) were in Goeppert Mayer's list of 'magic numbers'. The only problem was the cost. Today, 1g of Ca-48 costs $200,000. An accelerator uses 0.5mg an hour. In the glory days of Berkeley, elements could be produced with only a few hours of bombardment. By the 1970s the new elements had cross sections so low an experiment had to be run for weeks or months at a time just to get one atom. The cost of calcium soon mounted up. Even so, if Berkeley found the island of stability, it would have been worth every penny.

A day earlier, the Berkeley–Livermore team had attempted to make element 116. When the scientists had checked their catcher foils, they had seen a thin layer of black crud. Deciding to play it safe, the crud had been analysed along with everything else. Placing it in the detectors, it soon started pinging with spontaneous fission. They hadn't just made element 116, they had chemically isolated it too. Excitement gripped the lab. It was, potentially, the greatest discovery since fission.

Ghiorso picked up the phone and called Stanley Thompson to tell him the news. Aged only 64, Thompson was on his deathbed with cancer. The great chemist, a man who had come up with the technique to isolate plutonium and who had overseen the first atomic bomb's production in Hanford, was too ill to reply. '[I] talked to him,' Ghiorso later wrote in *The Transuranium People*, 'and was certain that Stan understood what had been accomplished.'

Before he settled for the night, Ghiorso checked his results. Something was nagging him. Taking a midnight stroll to the lab, he checked the filter paper, which should have been clean. It was giving off radioactivity too. Puzzled, he decided to get some sleep. The nightmare was to follow soon after.

In the morning, during his shower, Ghiorso realised what had happened. The Berkeley team hadn't seen a superheavy element at all. The radioactivity was just from fission, while the mysterious crud was the charred remains of glue used to cement the foils in place. It was a false alarm. But it was too late to tell Thompson. Shortly after his phone call with Ghiorso, he had passed away.*

In his eulogy, Thompson's son-in-law Kenneth Lincoln gave him a Lakota name: *Cante Ksapa* – 'Wise Heart'. 'He was a man to be liked and respected – a man of old values, the essential and simple ways of living … a man of good will, with many friends from all walks of life.' Glenn Seaborg's own tribute was just as potent. 'His radiochemical research during the Second World War rivals in importance the isolation of radium by Marie and Pierre Curie, and his leadership in the discovery of five transuranium elements must rank as among the leading chemical accomplishments of his time […] chemistry lost an extraordinary practitioner, and I lost a lifelong friend.'

Stanley Thompson was the first of the transuranic element giants to fall. But a new generation of researchers were emerging in Germany – and a host of new elements were about to follow.

★ ★ ★

Wixhausen is an unfortunate name for a borough. When it was founded a little over 800 years ago, Wickenhusen just meant 'houses on the pond'. Gradually, the name has shifted and German slang has evolved. Today, to the embarrassment or amusement of the locals, 'wix' sounds like the word for … uhm … to pleasure oneself. No wonder the local laboratory is largely referred to by the nearest city instead: Darmstadt.

---

* The Hanford site, where Thompson worked, is now known as 'the most toxic place in America', contaminated with leaking nuclear waste that costs billions to clean up. In 2018 a bill was signed into state law approving compensation for Hanford workers for a range of conditions, including several types of cancer.

The GSI Helmholtz Centre for Heavy Ion Research (GSI for short) is on the edge of town. Heading out to the lab means cutting past car dealerships and auto body shops and then through open fields and thick woodland. Gradually, things begin to get stranger. Trees give way to lavish, glass-fronted conference centres. Road signs warn you of migrating hordes of toads. Parks are decorated with scattered remains of particle accelerators lovingly preserved as art installations. It's only then you realise you've arrived at one of the most advanced laboratories in Europe. Despite its apparent isolation, the locals know all about this jewel in the crown of German science; the last time it had an open day, 11,000 people turned up at the gates. Out back, still a building site with only a few of the tools in place, GSI is building FAIR – the Facility for Antiproton and Ion Research – that will see scientists from 50 countries partnering to smash things together, annihilate matter and try to discover the very origins of the universe. In another part of the lab, nuclear technicians pioneer targeted therapy beams for destroying cancers, their machines able to set the beam intensity to pass harmlessly through tissue and bone before irradiating a tumour with complete precision. GSI unlocks worlds and saves lives.

The lab is part of the West German economic miracle, evidence of the staggering speed with which the country rebuilt itself after the Second World War. By 1969 nuclear physics departments across Germany had been building research accelerators at such a rate that there were 20 small-to-medium machines across the country, all in competition. This was such a duplication of effort that the universities in Darmstadt, Frankfurt and Marburg decided to pool their resources together; later, universities in Giessen, Heidelberg and Mainz joined the project. The result was GSI – the first lab that could not only compete with Berkeley and Dubna, but beat them.

'It wasn't clear where to build it,' remembers Gottfried Münzenberg, as we sit down in the cafe to relax with a coffee. 'There was competition with Heidelberg, with Karlsruhe … the idea was to come here because Darmstadt gave the land, and it's not far from Frankfurt Airport.'

Münzenberg is easy to interview. He cuts a relaxed figure: warm, affable and friendly. White-haired and bearing a passing resemblance to David McCallum's character Ducky from *NCIS,* he relishes in sharing his stories between pastry bites. Münzenberg hails from Wolfsburg – the home of Al Ghiorso's Volkswagen Beetle – and learned his English in the 1960s when he came to the UK to study at another car manufacturing hub: Luton. He chuckles as I assure him it hasn't changed much.

Once GSI's site was decided, the Germans had to build their accelerator. 'The idea was to create an accelerator that could accelerate all the elements,' Münzenberg recalls. 'It wasn't clear what to shoot … all that was clear was that the Berkeley recipes wouldn't work! So, we had to build an accelerator, and we had to build a spectrometer that was capable of detecting *everything.*'

It was ambition on a scale unmatched in nuclear science. Once again, it was a team starting from scratch and inventing their own way. Münzenberg was there from the start as an expert in ion optics, a fresh-faced postdoc with boundless energy who would do whatever was asked. Günter Herrmann, a physicist from the University of Mainz and another member of the GSI team, called in favours to get special permission to use radioactive materials for research ('You can't do that these days, it's forbidden,' Münzenberg says, 'but once we had permission, we had permission…'). Others worked on the target and the interactions with matter. The ion source, in a surprising twist, was given to the Germans by Georgy Flerov.

The German team applied themselves, but it wasn't all work and no play. Part of preparing the machine required the condenser to be cleaned with alcohol. Münzenberg and another colleague doctored the solution with a few 'organic contaminants'. It wasn't enough to ruin the clean-up, just enough to make the whole lab smell of whiskey. 'We had a lot of fun,' Münzenberg winks.

The man put in charge overall was Peter Armbruster. Handsome, dark-haired and sporting long sideburns, Armbruster had grown up in Dachau, under the shadow of its concentration camp. In the 1950s he had studied physics at the technical universities of Stuttgart and Munich and had become increasingly interested in heavy ion fission. As GSI's senior scientist, he had complete authority to decide what the German lab would investigate. 'Armbruster was *the* person,' Münzenberg recalls. 'It was the perfect team. Armbruster was the boss – never the group leader, always a director. He chose the [equipment]. No committee, no discussion. He prepared the whole thing … we did presentations, he wrote them up overnight and submitted them the next day.'

It was a sign of how fast things at GSI would move. The lab went from breaking ground to building the most advanced accelerator in the world – the Universal Linear Accelerator (UNILAC) – in five years. 'It was very exciting,' Münzenberg laughs. 'We started one year, made the design the next, the year after we built it … everyone said "what you do will never work" … but it was very fast. And we were lucky. Nature was good to us.'

Münzenberg is being modest. Under his leadership, the Germans were about to double the number of superheavy elements.

* * *

'So, this is the birthplace of our heavy elements.' Michael Block is the current superheavy element physics lead at GSI. He's giving me a walking tour, starting with the staggering vastness of UNILAC. We're in a long concrete tunnel, straight as an arrow, with a giant purple tube towering next to us as it runs the entire 120m (400ft) to the target. Its bulk is large enough for several people to climb inside and walk upright. At key points, latched into the gargantuan accelerator like suction cups on an udder, metal cylinders prod out. These are connected to thick black wires that eventually coil out into the wall.

A cyclotron is impressive because it's basically an alien disc lodged under the largest magnet you've ever seen. A linear accelerator is even more awesome. Watching it stretch out into

Figure 10 Maintaining the interior of the UNILAC linear accelerator, GSI Darmstadt.

the distance, it looks and feels like a sci-fi space cannon. GSI believes that it's also the better option for element hunting. 'You need a high-intensity beam,' Block says. 'And if you want high intensity, then you better go straight.'

We walk the tunnel, side by side next to the giant ion gun barrel, occasionally stopping to see where a part has been removed for maintenance. At these points the accelerator's hollow insides are revealed. Each section is a segmented chamber coated with gleaming copper. Inside, a line of metal rings like thick doughnuts are suspended by rods in the centre of the pipe's maw. The beam passes through these doughnut holes using the carrot-and-stick method perfected by 80 years of particle bombardment. At the start of the accelerator, the beam element (let's say uranium) is placed in a gas-filled chamber. A voltage is applied, tearing off a couple of the atom's electrons, before an electric field pulls the uranium ions into the accelerator. Here, pushed and pulled by changing the electric voltage at exactly the right moment, they hit their target at 30,000km (18,600 miles) a second – about 10 per cent the speed of light. The whole trip takes 10 microseconds.

GSI's system is designed to run several different experiments at once. Rather than firing a continuous stream of ions, the beam comes in 5-millisecond pulses every 20 milliseconds. About 1 per cent of ions are diverted to GSI's synchrotron, a closed ring where the ions continue to build up speed, flying around in circles hundreds of thousands of times before eventually hitting 270,000km (170,000 miles) a second. At that speed, you go from the Earth to the Sun in about nine minutes. Once FAIR is operational, its 1.1km (0.7 mile) accelerator ring will be able to speed up 500 billion uranium atoms to 95 per cent the speed of light in a single burst.

The other 99 per cent of the ions go on to the bitter end, eventually slamming into a piece of target that's thinner than kitchen foil. Block picks up an example, a strip of lead-208 that's already been bombarded. Its skin has flared deep yellow. 'See this different-coloured part? We don't use just lead – we use lead sulfide. It can take more beam before it's used up. Usually, it's good for running for a couple of weeks.' As lead is in easy supply (as is bismuth, the other possible target), GSI buys in bulk and mounts several targets at once on a rapidly spinning wheel: that way, the beam gets shared around rather than just hitting the same spot endlessly.

'In the good old days, we could run up to 6,000 hours of beam a year,' Block says. 'Nowadays we have three or four months, with maintenance breaks in between. A lack of manpower restricts us because of the construction of FAIR.'

We head away from the accelerator and into a control room that looks like the bridge of the starship *Enterprise*. Consoles with little lights, endless dials, readouts and gauges, all set in panels painted a potent burned orange that hasn't been in style since the 1970s. 'It's a bit like *Star Trek*, y'know?' Block grins. 'This was how we used to have things set up. Nowadays it's had a major makeover. We can control everything far more easily: switching between elements, ions, beam lines, changing the pattern of the pulses, changing the intensity of the beam and changing the energy.' Everything is simpler too. When GSI first formed, the laboratory had an entire electronics shack for the experiments that was filled with equipment. Today, advances in computing and digital measurement means a single panel the size of a desktop computer can run the whole experiment.

But UNILAC is only half of the reason for GSI's success. It's all very well being able to create an element – you still have to spot it too. When GSI was formed, the reaction cross sections were so low that it was virtually impossible to find your creations among all the background noise of ions pinging everywhere. GSI's answer was the Separator for Heavy Ion Reaction Products – SHIP. The name wasn't an accident; remembering the sea of instability, the Germans were convinced they could sail off to find the new elements before their superpower competition.

Block leads me through a maze of walkways, crawlways, narrow gaps and locked doorways. We're heading into the very belly of the beast – the home of the separator. SHIP is a complex piece of kit, but the idea behind it is relatively simple. If a fusion product is created (by the ions hitting the target), the laws of physics mean it'll be travelling at a much slower velocity than the ions in the beam. With a complex system of magnets and electric fields, the high-velocity ions can be steered away into a dead end, allowing only the slower particles to head into the detectors. It's a technique called in-flight recoil separation.

Free from the beam, any fusion products pass through detectors without all the noise. Now you can measure everything you need to prove an element has been created. First, an atomic-sized speed trap – two foils about 30cm (12in) apart – records the time of flight (usually about one to two milliseconds). Then, an array of detectors waits to pick up any signs of alpha radiation. 'Alpha decay goes in all directions,' Block says. 'So, if you only had one detector, at most you'd see half of it. We send the particles into a box covered with detectors. Most particles hit the sides of the box or the stop detector. It gives you a coverage of 80 to 90 per cent.'

I blink and try and get my head around things. Block is playing with one of the magnets, holding up a spanner a few inches away and letting go, watching it catapult in mid-air and attach to the side. Fortunately, he's also spotted my confusion. 'So, production of a new element. New atom recoils out, goes through the filter [to remove the beam]. It implants into the wall. Then you have a chance to register alpha decays.' Simple, right?

This isn't the only thing GSI can do. Gone are the guessing games about what you've made that are associated with spontaneous fission and alpha decay: GSI's equipment can weigh individual atoms as they fly past. As each isotope has a unique weight, this allows you to tell *exactly* what you've created. More importantly, mass measurement allows you to detect if you've made a single atom of something within the island of stability. While these atoms are unlikely to alpha decay or fission in the detector (that should take millions of years, remember), with mass measurements you can still spot if something has been created.

'How sensitive is your mass measurement?' I ask.

Block thinks for a moment, staring up at the ceiling and sucking in a breath as he comes up with an answer. 'Imagine a giant commercial airliner,' he finally decides. 'An Airbus A380. Our weight changes are so sensitive I could detect if you left a 1 cent coin on a seat in first class.'

All of these different measurements make the sensitivity of GSI's machine astonishing. 'We've hit a cross section of 90 femtobarns.' Block says. That's $10^{-41}m^2$. I try and keep in my head that cross section is a measure of probability, rather than size. At that level, even the most improbable reactions – the slimmest chances of fusion occurring – will eventually happen just through chance. All you need to do is keep your machine running long enough.

I think about the elements yet undiscovered. 'So, ignoring how much it would cost, if you kept everything running non-stop for 10 years, you might find something?'

Block shrugs. 'You can find anything if you have enough time.'

'And money,' I add.

Block nods ruefully. 'And money.'

★ ★ ★

Sigurd Hofmann is a man who comes prepared. Some interviewees sit down at a table and remember the good times; others tell you their own perspective then dare you to challenge their views. Hofmann has brought along a slide presentation and 26 peer-reviewed scientific papers. In the comfort of GSI's modern office block, we sit down over tea and biscuits and begin to delve into the world of element-making.

Hofmann was another early arrival at GSI. He was born in 1944 in what was then the Sudetenland of Germany, where his father worked making Bohemian glass for lamps. When he was 16 months old, the Russian army pushed into the region and the victorious Czechs forced his family to flee, first to Thuringia, then to Darmstadt. Here they settled, inadvertently placing the young Sigurd in one of West Germany's leading science hubs. Electing to stay close to home, he studied physics at the city's technical university, where he gained experience in nuclear reactions and the emerging science of computer programming. 'I had one of the first computers,' he recalls. 'It had 8k of memory, and I wrote a program so I could analyse gamma spectra.' When GSI was created nearby, the team needed a computer expert for SHIP. Hofmann was the perfect candidate.

The accelerator at GSI started up for the first time in 1976. It was the only lab in the world capable of accelerating uranium, and so for five years mostly shot uranium beams (why waste the best accelerator in the world on something others could do?). But the team yearned to try to get involved in element discovery. In 1976/77, the team managed to muster five and a half days of beam time to spend hunting for superheavies, aiming to create elements as high as 122. They had about as much luck as every other lab. 'I no longer remember in detail our great disappointment,' Hofmann wrote in his book *On Beyond Uranium*. 'Disappointments are easy to repress. But the records do not refer to any bottles of champagne.'

Then the SHIP team decided to try cold fusion. In the time since Oganessian's discovery, cold fusion had seemed dead in the water. Flerov had no interest in it and insisted the Dubna team went back to their previous scientific programme. Likewise, Berkeley remained unconvinced and still focused on Al Ghiorso's mantra: *1 alpha is worth 1,000 spontaneous fissions.* Neither decision is as ridiculous as it may seem in hindsight – Oganessian's tests had largely been a proof of concept. Any reliable discovery would require a particle accelerator that was capable of firing beams of ions far heavier than anything the Americans or Russians could accomplish and a detector more advanced than anything else in the world.

Fortunately, that's exactly what GSI had.

Oganessian remembers the Germans attempting to lure him to GSI, offering him a chance to run his beloved cold fusion and even bring his entire team over. He refused and remained in Dubna, though he couldn't resist sneaking in cold fusion experiments whenever he could.

Heinz Gäggeler, on the other hand, decided to take the Germans up on their offer. He had seen cold fusion in action under Oganessian and knew the Germans were the only team capable of pulling it off. 'It wasn't like they were stealing [the idea of] cold fusion,' Gäggeler stresses. 'If you want to produce elements with cold fusion, you need excellent beams and excellent detectors. At that time, the technology in the Soviet Union was lower than in the West. GSI had SHIP, but even so they still had a hard time. Three years without success … although, these days, nobody talks about that part.'

The three years came and went. They were not wasted. The team were learning, working out how to measure some of the rarest events in the world, which in turn created the rarest elements in the world.

Then in one week, from 12 to 17 February 1981, Münzenberg's team bombarded titanium into a bismuth target. Sure enough, the two nuclei collided, used that kick to sneak through the Coulomb barrier, then discarded a single neutron to reduce the chance of fission. The reaction produced the still-disputed element 105, before alpha-decaying into element 103.

Cold fusion worked.

A week later, on the morning of 24 February, the team tried a chromium beam (two protons heavier) into the target. At 10.48 a.m. they produced something. It followed the chain they had previously detected perfectly: first alpha-decaying into element 105, then into element 103. Nine hours later, they succeeded again; and again; and again. In four days, they had multiple perfect, pristine alpha decay chains, some even going down as far as californium before breaking apart. The atom's mass was confirmed by its time of flight and its energy as it landed on the detector. 'Within one week, we measured six decay chains,' Hofmann recalls. 'After this, Seaborg and the Berkeley people became interested. Flerov from Dubna too. They came to Darmstadt to visit our experiment.'

Seaborg arrived in September, Flerov in December. The two made an impression on Hofmann. Seaborg was 'a great man, tall and serious' … Flerov was 'also a great man but younger, not so tall and not nearly so serious.' Both giants of the superheavy world were impressed by the GSI team. Cold fusion didn't produce a lot of atoms (by 1988 only 38 atoms had been created) or isotopes with very long half-lives. It didn't matter. For the first time since 1955, everyone agreed that an element had been discovered and who had discovered it first. The Germans had created element 107.

GSI had gone from a clearing in the woods to the leading heavy element lab in the world in six years. Ghiorso flew to Germany with curium, hoping to make element 116. The experiment was a failure. So too was an attempt at Berkeley, with the GSI team flying out to join the Americans. Münzenberg loved Ghiorso, but the two tinkerers disagreed on the best way to proceed. Ghiorso wanted to leapfrog the unstable elements and head straight for the island of stability. The Germans, meanwhile, pointed out that this would have been impossible to prove. An alpha decay chain, for example, would have to pass through five unknown elements – each with unknown half-lives or chances of fission – before it joined up with known decay chains. Even if they succeeded, it would take years, perhaps decades, to prove the claim.

Instead, Münzenberg opted for another cold fusion experiment, this time looking for element 109. The reason for ignoring 108 was simple: even-numbered elements were more prone to fission. By leapfrogging, the team had a far better chance of creating one of the beautiful alpha decay chains that showed everyone exactly what had been made.

The new beam was made from iron-58 – a rare, expensive isotope that only makes up around 0.28 per cent of iron in nature. In August 1982 the team started the experiment. After five days they had the first sign of 109. After another 10 days, they had another. The team had been lucky – the hit came barely two minutes after they noticed they had run out of disk space and changed computer files. Two hits weren't enough to prove the discovery conclusively, but GSI were on a roll.

'In 1984,' Hofmann recalls, 'we studied iron plus lead (element 26 into element 82) and observed three decay chains, which decayed into the chain we'd previously studied. It was clear it was 108.' This wasn't expected: everyone suspected the isotopes would fission away rather than go through alpha decay. 'It was a sensation,' Armbruster and Münzenberg would later write in the *European Physical Journal H*, 'which became even bigger in the following experiment when we produced the [...] isotope 264 of element 108.'

As new experiments demystified the complex mechanics of nuclear forces and discovered yet more isotopes and decay chains, the GSI results were confirmed. By 1989, when Münzenberg was promoted to director and Hofmann took over the team, the Germans had irrefutable claims to elements 107 to 109.

The Germans were in the driving seat – and everyone else was scrambling to catch up.

CHAPTER FOURTEEN

Changing the Rules

In 1980 Glenn Seaborg accomplished the alchemist's dream of turning a heavy metal into gold. The Berkeley accelerator was loaded with bismuth foil (lead's neighbour on the periodic table – bismuth only has one stable isotope, so it's easier to separate), which was soon pelted with carbon and neon ions. The beams chipped off protons and neutrons, leaving scattered fragments of gold dust. Seaborg had become the modern King Midas. It was a stunt: the experiment cost a day of beam time (worth about $120,000) to make quantities so minute they could only be detected through their radioactive decay. 'It would cost more than one quadrillion dollars per ounce to produce gold by this experiment,' Seaborg reported to the Associated Press. Even so, he had shown once and for all that science had surpassed alchemy.

In 1984, just as GSI was discovering element 108, another miracle happened: Darleane Hoffman broke her 40-year-old vow and agreed to become a teacher. Stepping away from Los Alamos, she took up the position of tenured professor at Berkeley, and with it the leadership of the heavy element group. Ghiorso happily folded under her wing, bringing his endless enthusiasm (sometimes, a little too much enthusiasm) to whatever their latest project happened to be. The three US superstars – Hoffman, Ghiorso and Seaborg – were all in one place, bringing a combined 125 years of heavy element experience to the table.

But the climate in the US for element hunting soon changed. On 26 April 1986 a late-night safety test at the Chernobyl power plant, near the town of Pripyat on the northern frontier of Ukraine, went horribly wrong. At 1.23 a.m. the plant's number-four reactor exploded, shooting its parts through the roof of the building and starting fires across the complex. Almost 7t of radioactive matter scattered into the atmosphere.

The Chernobyl accident shifted US public opinion of nuclear power. Already, a less serious accident at the Three Mile Island Nuclear Generating Station in Pennsylvania on 28 March 1979 had made the public wary; Chernobyl tilted warped perceptions into outright fear. Nuclear opposition cited the 'China syndrome': that if a core went into meltdown, it wouldn't stop tunnelling its way through the Earth until it reached China.* The climate that had emerged in the 1940s and 1950s, a world excited at the prospect of new, clean and boundless energy, was over. Nuclear power was a pariah.

For Seaborg, the scepticism was a blow that almost unravelled his lifetime's work. 'I can't claim to be blameless,' he wrote in his autobiography. 'My early boosterism of nuclear power may have contributed to later problems [...] plants were prematurely escalated in size to proportions that strained the technology and magnified the potential consequences of an accident, no matter how unlikely.'

As political will faded and budgets shrank, the entire science of radiochemistry came under threat. It was a problem only compounded by the veil of Cold War secrecy. Sometimes, information couldn't be published because of national security. On other occasions, the culture of mistrust and caution meant details that could and should have been shared were kept under wraps. There was also a lack of new blood in the labs. As the older element hunters refused to retire, the pipeline of younger researchers was blocked, limiting opportunities for recruitment. A few brilliant candidates found placements; many others looked elsewhere for their careers.

Funding cuts were even more devastating. Berkeley couldn't call on the deep pockets of the government any more. The team asked for a rare nickel isotope for their beam to hunt for element 110; they were told the money wasn't available. They asked for a separator as good as the

* It wouldn't, but since when did facts get in the way of a good soundbite?

German SHIP, but funds went elsewhere. Eventually they had to build the machine themselves from spare parts – its emergency valve, designed by Ghiorso's son Bill, was made from the spring of a rat trap. Ghiorso, never one to pass up a good name, called it the Small Angle Separator System (SASSY).

Despite its make-do innovations, SASSY had no new elements to separate. Super-HILAC was hooked up to the bevatron, the giant velodrome-sized accelerator that dominated the hill, for use as an injector. The new combined monster – originally Ghiorso's brainchild – meant other groups whose projects had more immediate applications sucked up every ounce of funding and beam time. On the rare occasions Super-HILAC was available, misfortune would inevitably strike. On one Friday, with the GSI team visiting, the entire lab was suddenly plunged into darkness. The local power company had an agreement with Berkeley Lab to cut the lab's power if they were using too much juice – and Ghiorso's machine had just pushed past the limits. The Americans and Germans scrambled about in the dark, talking over each other in complete confusion. It was such chaos that everyone forgot to turn off the helium flow to the separator. In an almost exact repeat of the 1959 HILAC explosion, the helium pressure rose and punched through the entrance window, breaking the curium target.

Fortunately, instead of showering the lab, the curium just contaminated SASSY. 'Poor Ghiorso,' Sigurd Hofmann later wrote in *On Beyond Uranium*, 'he now had to spend the weekend cleaning up the mess. The rest of us felt guilty but left him to it; without special permission we were not allowed to help.'

★ ★ ★

If things were tough for the Americans, in Dubna Flerov's group were feeling the effects of the Soviet Union crumbling around them. As the Cold War funds dried up and the Eastern Bloc began to crack, the JINR scientists started to see their funds wither, dwindle and die away. If they wanted to continue their work, they would have to do something drastic.

In 1989 Flerov was attending one of the regular superheavy conferences that saw the scientific community come together. Until that point, any Russian collaboration with Americans had been through private universities or formal visits (even if they occasionally broke down into Johnny Rivers jive sessions). Flerov decided that this would end. He spoke with Ken Hulet, the head of the delegation from Livermore, and suggested they join forces.

The two men chatted at length. Dubna had a cyclotron. Livermore had targets to use and expertise in detectors and equipment. Unspoken, but equally important, Livermore also had a level of scientific credibility; while the Russians were brilliant scientists, the battles with Berkeley had cast a shadow over their accomplishments. If Livermore came on board, Berkeley couldn't question the validity of their experiments any more.

Hulet and Flerov, chemist and physicist, ended their chat with a handshake. It was unprecedented. The Americans and Russians had bridged the Cold War divide. Dubna and Livermore – a Soviet lab and a US nuclear weapons facility – were partners.*

A few months after Flerov's pact with Hulet, the Berlin Wall fell. Early the next year, the first Livermore researchers, Ken Moody and Ron Lougheed, arrived in Dubna. There were the usual Cold War hassles – only one long-distance phone call permitted, KGB wiretaps in the hotel and minders around town – but once the team passed into the lab, there was only science.

Georgy Flerov had witnessed the history of nuclear physics writ large in technology, warfare and diplomacy. Finally, as the shadows of a silent war between superpowers fell away, he had masterminded the coming together of the US and USSR

* Not everyone supported the collaboration. Matti Leino was invited too, but felt Flerov's way of working did not suit his style.

as a team. It was probably his greatest accomplishment. He would never see its results. In November 1990 he died suddenly in Moscow, aged 77. In Dubna all work stopped for three days as the team reacted to the loss. Moody and Lougheed were still in the USSR at the time, and attended the numerous memorials and services held in Flerov's honour. The collaboration had lost its leader, but the work would continue.

Flerov also missed the concluding phase of the transfermium wars. In 1986, at the request of the Germans, the governing bodies of chemistry and physics, IUPAC and IUPAP, assembled a working group to settle the debate surrounding the superheavy elements. Known as the Transfermium Working Group (TWG), its members had to define when an element counted as being made, and who had got there first. It was like playing the first-ever game of football, only for the referee to explain the rules and work out who scored first after the match had ended.

Science was changing its rules. Yes. It can do that.

★ ★ ★

For the past 120 years everything in the world has slowly been getting slightly lighter. It's because a lump of metal in a Parisian suburb wasn't doing its job properly.

For millennia, there had been no need for everyone to have exact weights – nothing in the world required that level of precision. But by the end of the Victorian era, this was beginning to cause problems. Measurements were becoming so exact that it became essential to have a set standard for mass.* In 1889 scientists agreed to define a kilogram as the mass of

* A similar process happened with time. Ship's captains needed the exact time to establish their position at sea, which drove the British Admiralty to develop accurate clocks. The advent of railways also made a set time essential – before then, nobody was able to travel quickly enough to notice that London was four minutes behind Cambridge or two minutes ahead of Southampton.

Le Grand K, a cylinder of platinum-iridium alloy at the International Bureau of Weights and Measures in Sèvres. The mass of *Le Grand K* would always equal *exactly* 1kg (2.2lb): no more, no less. Once every 40 years, the weight was taken from the vault and used to verify the mass of 67 copies stored around the world. From there, the mass of everything, from your bathroom scales to your grocery shopping, was set accordingly.

Le Grand K is stored at constant humidity and temperature next to six accompanying kilogram weights in a triple-locked vault. Security is so tight that one of the necessary keys to access it is usually kept abroad. Until recently, the reason for this pantomime lockdown was simple: if *Le Grand K* changed (for example, if someone cut a bit off), then the definition of what counts as a kilogram would suddenly shift. It may sound like the madcap ploy of a James Bond villain, but you can't have someone monkeying around with the weight of the world.

Fiendish plots aside, the real problem is that *Le Grand K* has been gaining mass for decades. Even in its carefully controlled vault, tiny air pollutants – invisible specks of dust – can settle on the platinum, making the lump slightly heavier. This meant the definition of the kilogram became heavier, so everything else in the world officially became lighter.

This couldn't go on, so in 2010 the International Bureau of Weights and Measures decided to change the rules. Rather than define a kilogram based on *Le Grand K*, the bureau agreed to base the weight of a kilogram on the Planck constant – a fixed number central to modern quantum physics.*

---

*The obvious question is 'How do you measure the Planck constant?' Scientists did it in two ways. First, they used a very precise device called a Kibble balance, which measures weight using electric current and voltage. Second, they made a series of 93mm silicon spheres – the most perfectly circular objects ever created – which contained a known number of atoms, and which also allowed them to eventually work out the Planck constant. Fortunately, the numbers agreed: making the definition of the Planck constant – and therefore the kilogram – accurate to a few parts per billion.

Bizarrely, that also links the kilogram to the definition of the metre and the second – but at least it means we don't need to worry about the mass of the world shifting if something happened to *Le Grand K*. By the time you read this, the switchover has probably happened (it was scheduled to take place on 20 May 2019).

These changes to things we think of as basic rules of the universe – when was the last time you doubted if the kilogram was real? – aren't just idle tinkering. As science evolves, we always need better definitions. For the Victorians, *Le Grand K* was enough; today we're at such a level of precision that it just won't do.

By the late 1980s the same thing had happened with element discovery. For Antoine Lavoisier in the eighteenth century, an element was something that couldn't be simplified; for Ernest Rutherford, it had been defined by having a unique number of protons in its nucleus. But nuclear science teams were already playing around with quasi-fission, making elements that *almost* came together before breaking apart. Did that count as a new element? And what proof did you need that anything had happened at all?

In the meantime, everyone else had a headache when trying to talk to each other about elements. If someone said 'rutherfordium', for example, did they mean the Russian name for element 103, or the American name for element 104? The situation had become so fraught that IUPAC even introduced 'placeholder' names for the elements based on their atomic number. Element 104 became 'unnilquadium' (one-zero-four-ium); 112 was 'ununbium' (one-one-two-ium). 'In this Cold War, one of your strongest weapons in the public relations battle was the name you proposed for "your" element,' explained Norman Holden, the man who invented the system, in the magazine *Chemistry International*. 'You would never give up your strongest weapon and accept a neutral name. This would indicate that you didn't believe strongly in your scientific case for the right to discovery.'

This was the challenge facing the TWG. For three decades, both the US and USSR had published papers that varied in quality. Some were flat-out wrong; others contained flashes of scientific brilliance the other team refused to admit were correct. Much of the bad data hadn't been retracted, while much of the good was hidden away in private lab records the other side couldn't access.*

The TWG is largely considered to have been flawed from its inception. To remain neutral, IUPAC and IUPAP decided to avoid including anyone who could be biased either way. Unfortunately, element discovery has such a small community that *everybody* had a viewpoint. The result was a group of eminent scientists from just outside the superheavy field fronted by the UK's Denys Wilkinson – an outstanding nuclear physicist, but someone who had no superheavy element or radiochemical expertise.

The TWG set off around the world, visiting each of the labs who had a solid claim to a new element. Politics dogged it at every turn. Originally, Dubna was due to be visited before Berkeley; in a last-minute decision that infuriated Ghiorso and Hoffman, the TWG decided to visit Dubna last. The Americans felt swindled, accusing the Russians of dirty tricks by ensuring they had 'the last word' and the 'the extraordinary and highly questionable advantage of having the TWG "collaborate" with them in "retrospective re-evaluation" of the data'.

Seaborg and Yuri Oganessian started a series of back-channel discussions to try and find a compromise before the TWG forced their hand. Both Berkeley and JINR were content to agree that they had both at least contributed to the discovery of the superheavy elements. The sticking point was element 104. The Russians wanted 'kurchatovium', after Flerov's mentor. The Americans wanted anything *but*

---

* The TWG wasn't the first attempt to unpick the transfermium wars. In 1974 a neutral committee – plus three members from the US and USSR – was asked to decide who had discovered which element. Ahead of its first meeting, the chair had asked both sides to put forward their case. For the US, Darleane Hoffman recommended that all elements be awarded to Berkeley; the USSR team walked away. The committee never held a meeting.

'kurchatovium'. A letter Seaborg and Ghiorso wrote at the time to GSI's Peter Armbruster underlines their feelings: 'We would NEVER [their emphasis] agree to the naming of an element after Kurchatov,' Ghiorso and Seaborg fumed, 'any more than we would to the naming of an element after an American inventor of the hydrogen bomb!'

It was a complete impasse. All either side could do was wait for the TWG ruling and hope it recognised their claims and right to name the elements. In 1991 the TWG announced its findings. The first half of its remit (what constituted a new element) was relatively uncontroversial: 'The experimental demonstration, beyond reasonable doubt, of the existence of a nuclide with an atomic number not identified before, existing for at least 10^{-14} seconds.' There was muffled grumbling about what 'beyond reasonable doubt' meant, but otherwise the rule made sense: the required time was how long it took for the positively charged nucleus to attract negatively charged electrons, forming an atom. Fine.

Next, the TWG announced who had created each element. In a controversial move, a few of the elements were even shared between groups, the TWG reasoning that both had played a significant role in the discovery. The ruling was as follows:

101	Berkeley
102	JINR
103	Berkeley and JINR (joint)
104	Berkeley and JINR (joint)
105	Berkeley and JINR (joint)
106	Berkeley and Livermore
107	GSI*
108	GSI
109	GSI

* GSI is recognised as having priority for element 107, although Oganessian was invited to be the 'godfather' of the element because of his contribution to its discovery.

The decision satisfied no one. The Russians and Americans both felt their work was not recognised; the Germans felt caught in the middle; and the Swedish team's claim to element 102 had been dismissed completely.

The attacks began almost immediately. The Berkeley team attacked Dubna's right to element 104, insisting that 'acceptance of this conclusion would be a disservice to the scientific community' (Denys Wilkinson didn't take the Berkeley criticism lying down, insisting the reply 'sits ill with Messrs Ghiorso and Seaborg' and refusing to change the report).

Eventually, the Americans backed down; once again, political pressure was mounting to settle the argument for the good of international relations. The US team wrote to accept the TWG proposals – although Matti Nurmia was excluded from the letter. 'There was an agreement signed by everyone except myself,' he recalls. 'I was left out of it because I was known to be so critical of the Russian work I would not have agreed to the conclusions.'

But the real problem remained: elements 103 to 105. The TWG had given two groups, already using two different terms, the equal right to choose the elements' final names.

It was asking for trouble.

CHAPTER FIFTEEN

How to Name Your Element

In 1993 Al Ghiorso received a phone call from the *New York Times*. Fifty years earlier, his only claim to fame had been illegally breaking a ham radio record. Thanks to the TWG ruling, he was officially credited with the discovery of 11 elements. This had beaten a 185-year-old record for element discovery held by the British chemist Humphry Davy. Albert Ghiorso, the bootlegger's son with a bachelor's degree, was the most successful element hunter of all time.

For Ghiorso, the sweetest triumph was recognition that Berkeley had discovered element 106. It had taken 20 years, but finally the team could pick a name. Ghiorso had been thinking of 'alvarezium', after his friend and colleague Luis Alvarez. Other inspirations included Frédéric Joliot-Curie (mainly because the Russians had been considering it for element 102); famous figures of the past such as Isaac Newton, Leonardo da Vinci or Christopher Columbus; and the mythical sailor Odysseus. Matti Nurmia had pressed for an alternative: finlandium. 'At the time,' Nurmia reasoned, 'there were two Americans and three Finns in the group ...'

The names were still rumbling in Ghiorso's mind as he picked up the phone. The reporter on the line was Malcolm Browne. Browne had started out life as a chemist but had turned to journalism after being drafted in the Korean War and assigned to the military newspaper *Stars and Stripes*. He'd risen steadily and was eventually appointed chief correspondent for Indochina for the Associated Press. There, he had won the Pulitzer Prize for his haunting image of the death of Thích Quảng Đức, a Buddhist monk who had set himself on fire in an act of protest. Since 1977 Browne had mixed war reportage with the science beat. When it came to element discovery, there wasn't much difference.

'What are you going to name element 106?' Browne asked, mischievously probing for a scoop. 'Ghiorsium?'

Ghiorso laughed. He'd first heard the suggestion in 1957, at a Christmas party, when Glenn Seaborg had given him a large bottle labelled: '110 – Ghiorsium: a worthless metal … can only be prepared between the hours of midnight and 6 a.m. … spontaneously inflammable … just generally falls apart in a hell of a hurry. Also has an automatic transmission.' Ghiorso didn't share the anecdote with Browne. Instead, he deflected the question, chatted amiably for a while, then put the phone down.

But Browne's comment had sparked something in Ghiorso's mind. No element in history had ever been named after a living person. However, there was no rule against it – just an unofficial chat with the Russians that such an idea would be avoided. Still, Ghiorso had an idea, and told the rest of the discovery team. They all agreed. On 2 December 1993 Ghiorso created a special cover for his folder charting the story of element 106 and made his way to Glenn Seaborg's office. At Ghiorso's prompt, his friend and colleague of 50 years opened it and read the inscription.

Dear Glenn, the team has decided unanimously that the only name for element 106 is yours!

Seaborg was astonished. 'I was incredibly touched,' he later recalled in his autobiography. 'This honour would be much greater than any prize or award because it is forever; it would last as long as there are periodic tables.' A name scrawled into an 1867 ledger by an immigration officer who couldn't spell Sjöberg was about to grace science's chemical trophy cabinet. Element 106. Seaborgium.

By the time Browne and Ghiorso had their chat, the other missing element names had started to slip into place. Element 101 was, indisputably, 'mendelevium'. Element 102 was more or less accepted as 'nobelium', despite both the US and Russians agreeing that the Swedish team who'd come up with the name hadn't discovered the element at all. Element 103 was 'lawrencium'.

The next two elements, 104 and 105, were harder to pinpoint. The Americans were adamant they wanted 'rutherfordium' and 'hahnium'. The Russians, meanwhile, insisted on their right to name the elements as 'not only an honorary privilege for the discoverers but also an acknowledgement of their intellectual property and the expenses of the laboratory'. Their names for 104 and 105 remained 'kurchatovium' and 'nielsbohrium'; maybe 'dubnium' as a potential compromise.

That left GSI. Their response to the TWG was the most magnanimous, acknowledging the contributions of both the Russians and Americans. Even so, Gottfried Münzenberg remembers pressure coming from both sides. 'I got phone calls from Berkeley in the middle of the night,' he told me. 'They were saying things like "We're sitting here with Seaborg and we want to propose names." They didn't want Kurchatov, but they did want Seaborg! It wasn't consistent. When we announced our names, we got a phone call from Berkeley saying they wouldn't come [and support our name choices] if we backed "kurchatovium".'

The Germans tried to stay neutral. They backed 'seaborgium', but suggested element 107 should be 'nielsbohrium' (hoping that would ease tensions around element 105), and because they 'fully agree[d] with the Dubna group that Niels Bohr highly merit[ed] to be honoured by the name of an element'.*

Fortunately, their two remaining elements, 108 and 109, were easy to name without controversy. GSI was in the German state of Hesse, the Latin name for which was Hassia. Seeing as how everyone else named an element after their local area, they decided to stick to tradition and proposed that element 108 would be 'hassium'.

---

* You'll remember that 'niels' was added to distinguish the element from boron (which, in German, is *Bor*). There was another good reason too: Niels's son Aage Bohr had won the Nobel Prize in 1975 – the first name settled any confusion about which Bohr they were talking about.

Münzenberg's choice for element 109 was intended to right a past mistake. The most important German contribution to nuclear science had been the discovery of fission. Berkeley were already proposing an element after one leader of the team that had made the breakthrough, Otto Hahn. The Germans wanted the other main contributor remembered too. Lise Meitner had been overlooked because of rampant sexism in the Nobel Committee; she had been persecuted and forced to flee her home by an evil that wanted her dead because of an accident of birth. Giving her an element was a chance for atonement.

'Meitnerium,' Hofmann told me, 'was a clear choice. She suffered so much as a scientist under the Nazis in the 1930s and had to escape Germany because she was Jewish. Despite this she was the main physicist in Otto Hahn's team – at a time when women were hardly accepted.' The GSI team couldn't put an end to sexism in science, but they could damn well remind people that great women existed. With curium named after both Marie and Pierre Curie, meitnerium would become, to date, the only element named solely after a non-mythological woman.*

The Germans hadn't waited around for the world to accept their elements – they had delayed for 10 years and weren't going to sit around any longer. On 7 September 1992 they held a ceremony and made the names official.

All the teams could do was wait to see if chemistry's IUPAC, and its physics equivalent IUPAP, would accept their choices.

★ ★ ★

Ian Fraser Kilmister isn't someone you'll find in a science textbook. As the leather-faced, raspy-voiced lead singer of rockers Motörhead, 'Lemmy' was better known for his trademark mutton chops, cowboy hat and a hard livin'

* Cerium, europium, niobium, selenium, tellurium and vanadium are all named, directly or indirectly, after goddesses.

lifestyle than his skill with test tubes and a Bunsen burner. I once saw him live, on a wet, miserable evening in Manchester, UK. He belted out his hit 'Ace of Spades' with a sullen growl, only to come to a stop at the second verse. Then, guitar hanging loose, he walked over to the amp controls to crank them up even higher before launching back where he'd left off. My hearing didn't recover for three days.

But in spite of his absent scientific pedigree, Lemmy also inspired the most popular suggestion for an element name in history. The singer had the fortune – or misfortune – of dying in late 2015, just as a new batch of elements were confirmed. The announcement inspired the public to get involved in suggesting possible names, and Lemmy's was at the top of the list. By the time the element's names were settled, a petition of 157,438 supporters demanded that element 115 be 'lemmium'. What better name could you have, they argued, for a *superheavy metal*?

Sadly, a new element could never have been named after Lemmy. After the chaos of the transfermium wars, IUPAC had put together some guidelines about just what could have an element named after it. In 2016 the recommendations became official. The categories were broad but defined. First, an element should end in '-ium' (unless it was a halogen, like chlorine or iodine, in which case it got an '-ine', or a noble gas, like neon or krypton, which earned itself an '-on'). Next, elements could only be named after five things: a scientist; a property of the element; a mineral the element came from; a place; or a mythical creature. This is basically just tradition – but the categories are broad enough to let scientists go wild if they wish. There is one final rule, designed to dispel any confusion: you cannot use a name already given to a chemical element, or that was once given and used widely before it was retracted.

The rules put paid to Enrico Fermi's 'ausonium' and 'hesperium' (they were widely used) and ended Lemmy fans' dreams of their hero appearing on the periodic table: while his status borders on legendary, Stoke-on-Trent's favourite rock god[*] doesn't qualify as a mythical creature. 'IUPAC has to be very neutral,' explains Lynn Soby, IUPAC's executive director. 'We just manage the process of working with the assigned laboratories and scientists. There are limits, guidelines if you will. But it's still broad if one thinks about mythological characters. There are lots of different creative opportunities for names.'

Today, the IUPAC process for recognising a new element is simple. A panel assesses all claims and decides which, if any, have enough evidence to prove an element has been made. The panel then announces which lab – or labs – have priority: the right to call themselves the discoverer. As the discoverer, the lab gets the honour of suggesting a name for the new element to IUPAC. If they don't do it within six months, they lose out and IUPAC has to name the element. In an effort to avoid a repeat of the transfermium wars, in the case of a joint discovery where the laboratories fail to agree on a name within this time period, the honour again defaults to IUPAC.

IUPAC doesn't have to approve the suggested name automatically: since 1947, the final decision has rested with IUPAC's council. In theory, it is possible for IUPAC to reject the name and put forward their own choice – although this, Soby muses, is a last resort.

Provisional name selected, the next part is the public consultation. This is basically a sense check to make sure the name isn't stupid or offensive to a global audience. 'One never knows what the general public will bring,' Soby says. 'One issue is that the name is going to be used in multiple languages. We need to check whether it has any negative connotations, whether it is pronounceable in all languages and whether it is sensitive in any way.' Just because a name seems harmless in English and French doesn't mean it isn't rude in Turkish. 'We want people to look at the proposed names, really evaluate them in their native language and see if there are any

---

[*] With apologies to Slash from Guns N' Roses, who is also from Stoke and comes a close second.

problems.' For example, given its dirty meaning in German, it's unlikely 'wixhausium' would be approved.

It's only after this public consultation that the element names are finally accepted; even then, it still takes a few months before IUPAC and IUPAP's joint working party ratifies the decision. Once that happens, the names are locked in forever – even if it's later shown that the element was discovered by someone else.

At least, that's what's supposed to happen. In 1994 IUPAC's announcement of the new names didn't go smoothly. The suggestions from Berkeley, Dubna and GSI had been gathered up, considered and put in front of a 20-strong panel of chemists from around the world. Faced with competing names from groups with equal priority, the working party decided to try and strike a compromise and blend the American, Russian and German choices together.

The new elements would be:

101	mendelevium
102	nobelium
103	lawrencium
104	dubnium
105	joliotium
106	rutherfordium
107	bohrium
108	hahnium
109	meitnerium

The new names were a terrible mistake. 'Joliotium' had never been used for element 105 before. Niels Bohr's first name had been hacked off element 107, which made it sound identical to boron, while 'hahnium' – an American suggestion – had been plastered on an element discovered indisputably by the Germans. Worse, 'rutherfordium' had previously been elements 103 and 104, but it was now sitting as element 106, its third place in the periodic table in 30 years. It soon emerged as the most controversial of IUPAC's decisions. The US team had announced their choice of 'seaborgium' to the world and it had been supported by the Germans. Instead of arguing, IUPAC had, retroactively, decided to create a new rule: elements couldn't be named after a living person. 'Seaborgium' was off the table. As *The Economist* noted at the time: 'When it comes to giving things names, scientists have a habit of throwing logic out the window.'

The superheavy community didn't take the decision lying down. The US National Academy of Sciences, IUPAC's biggest supporter, allegedly threatened to pull funding if IUPAC didn't back down on the names and allow 'seaborgium' to stand. 'I don't know what motivated IUPAC to do it,' recalls chemist Paul Karol. Today a member of the IUPAC/IUPAP joint working party in charge of deciding when an element is discovered, at the time Karol was so incensed he wrote a White Paper attacking the working party's choices. 'I can understand them floating the idea of not naming elements after living scientists, but it had become an edict. They didn't put it out for public review, so they immediately took heat. Seaborg was universally regarded as a giant in science, he'd made huge contributions. It was stupid.'

Karol felt the Americans were being portrayed unfairly as the bully boys of the element world while the Russians, who had somehow pressured IUPAC to rule out 'seaborgium', were getting what they wanted. Karol's suspicions were correct – although the Russians' motives were less political and more centred on the gentleman's agreement they had with Berkeley. 'It was agreed not to give names from living scientists by the people involved [in the discovery dispute, before they announced seaborgium],' Andrey Popeko recalls. 'It was agreed. From Dubna, we agreed to move kurchatovium. We appealed seaborgium because it was agreed! I have nothing against Glenn Seaborg. *But. It. Was. Agreed.*'*

---

* I couldn't confirm this agreement with the surviving members of the Berkeley team, but the emphasis here isn't for show – even 25 years on, the emotions stirred are felt acutely.

IUPAC suddenly found itself under assault. Buckling under pressure from their largest backer, the working party quickly dropped the 'no living scientist' rule and met again, creating another list of names in 1995. This time, to appease the Russians and ease any soreness about Seaborg's name appearing, they decided to add Georgy Flerov's name to the table and mostly adopted the Russian names:

101	mendelevium
102	flerovium
103	lawrencium
104	dubnium
105	joliotium
106	seaborgium
107	nielsbohrium
108	hahnium
109	meitnerium

Bohr had its 'niels' back, and 'rutherfordium' – previously three different elements – had vanished entirely. Still, the international community screamed for blood. The Americans had 'seaborgium' but felt 'all the rest of the USA-proposed names were being held hostage in return for retaining it'. Comments flew in from chemical societies around the world, with the Chinese and Japanese chemistry communities backing the Americans.

The chaos of the transfermium wars had reignited. In 1996 the Germans decided to hold a celebration for Peter Armbruster's sixty-fifth birthday and invited the Americans and Russians along. Gottfried Münzenberg told me what happened next: 'Armbruster had invited Seaborg, Ghiorso and Oganessian. They gave talks, and with Sigurd [Hofmann], we all came together in the evening to discuss the elements. We offered them wine. Ghiorso said: "We won't drink wine." So, we ordered sparkling water. Ghiorso said: "We don't want sparkling water." So, we brought them still water. Ghiorso said "We want water from the tap!" That was the kind of atmosphere ... we were neutral, we had *no interest* in the names, our aim was to find a solution.'*

In 1997, bloodied and weary by two years of threats, protests, arguments and complaints, IUPAC met in Geneva and put together its final list:

101	mendelevium
102	nobelium
103	lawrencium
104	rutherfordium
105	dubnium
106	seaborgium
107	bohrium
108	hassium
109	meitnerium

The Germans still didn't like losing the 'niels' from 'bohrium', but at least they were given the names they had suggested five years earlier. The Americans were mostly satisfied, although still insisted 105 should be called 'hahnium'. The Russians lost any reference to Kurchatov and Flerov; they were also forced to accept the old Swedish name for element 102. But this time there was no protest. The names were settled.

The choice nearly claimed one victim. Driving along the Californian coast, one of Seaborg's two living daughters heard the announcement of the new element names on the radio. She knew about the rule that elements couldn't be named after a living person but had missed that the decision had been reversed. When she heard her dad's name read out as an element, she came to the only natural conclusion: he was dead. Bursting into tears, she almost swerved off the road before she regained enough composure to pull over and call her (still alive) father.

The transfermium wars were, at last, over. In their span, they had seen three different elements named 'rutherfordium',

* Hofmann remembers it was Seaborg who demanded tap water, but you get the idea.

three different names for element 102 and two different versions of 'bohrium'. Finally, everyone could move on.

Because of the 'no repeating element names' rule, some of nuclear chemistry's pioneers would also never have an element named after them: Frédéric Joliot-Curie and Otto Hahn's names had been discarded forever. The latter was a poetic twist. Hahn had been remembered by the Nobel Committee when his partner Lise Meitner had been overlooked; now she would appear on the periodic table while he would be forgotten. Chemistry has an odd habit of reaching equilibrium.

For Glenn Seaborg, the story was different: 'seaborgium' was confirmed. It was the ultimate honour and, as his colleagues were quick to point out, made him the only person to whom you could address a letter using only chemical elements:

Seaborgium,
Lawrencium Berkelium,
Californium,
Americium.

Despite being 86 years old, Seaborg kept a schedule that would have exhausted a man half his age. He was in the process of writing two books, was still bending the ear of US presidents and had been hammering the cause for science in education, trying to convince California's governor to make learning a priority. In August 1998 Seaborg flew to Boston for the American Chemical Society's fall meeting: the biggest chemistry event on the planet. Over the course of a few days, almost 18,000 chemists descended on the city, picking it clean of hotel rooms, conference halls and those little lanyards to hold a name badge. Its delegates represented every discipline: materials science, agrochemicals, organic and inorganic chemistry, analytics, geochemistry, toxicology, medicinal chemistry and more.

There, in front of his peers, Seaborg collected a lifetime achievement award – something not given before or since. The 150,000 members of the American Chemical Society had voted Seaborg the third greatest chemist of the past 75 years.

The other two, Linus Pauling and Robert Burns Woodward, had died; in the eyes of his peers, that made Glenn Theodore Seaborg the greatest chemist alive.

Seaborg collected the award, gave a speech, then descended to walk the hall, signing the new periodic tables that bore his name. That night, as part of the routine he had started 60 years earlier, Seaborg decided to stretch his legs and walk up and down the hotel emergency stairs. There, alone, he suffered a stroke and collapsed. By the time help arrived, Seaborg was almost entirely paralysed. Six months later, on 25 February 1999, bedridden and suffering from arthritis so severe it made any movement agony, he chose to stop taking food and end his life.

With Seaborg, the era of the superheavy element giants passed into history. Of the original element hunters, only Al Ghiorso remained. But the race to the next elements had continued. Throughout the 1990s, Berkeley, GSI and the Dubna–Livermore teams had all continued to push boundaries a young Seaborg could barely have imagined.

On one of his last visits to his friend, Ghiorso even had to honour a $100 bet made decades earlier. It had been Glenn Seaborg's dream to see the shores of the island of stability. Under Oganessian, the Dubna–Livermore team had discovered a single atom of element 114. It should have been so unstable that it wouldn't even have met IUPAC's definition of an element. Instead, it had a half-life of 30 seconds.

'I wanted Glenn to know,' Ghiorso would later recall in *The Transuranium People*. 'I went to his bedside and told him. I thought I saw a gleam in his eye, but the next day, when I went to visit him, he didn't remember seeing me. As a scientist, he died when he had that stroke.'

The Dubna sighting of element 114 wasn't the only superheavy breakthrough of the 1990s. Even as the naming arguments raged, the hunt had continued – and both Dubna and GSI had been busy.

PART THREE

THE END OF CHEMISTRY

CHAPTER SIXTEEN

After the Wall Came Down

Life in Russia in the early 1990s was tough. In August 1991, the same year the TWG's first report emerged, there was a coup attempt to topple Mikhail Gorbachev. The Russian economy had gone into free fall; male life expectancy had been pegged back eight years.

In Dubna, Yuri Oganessian looked at his diminished staff, trying to work out what to do. Flerov, he conceded, would have known the answer; the old director always achieved his goals through sheer force of will. While the fortunes of other labs at JINR had ebbed and flowed, boomed and busted, Flerov had always found a way to keep the superheavy programme at the forefront of socialist science. But Flerov, the great master of the elements, was gone.

For JINR, the aftermath of the Soviet era had been devastating. The money vanished overnight, advanced research projects were frozen and the specialists who had gathered from across the world began to slip away to other jobs. Nobody blamed them. Life in the private sector was well paid and secure. Those who stayed weren't being paid at all.

Oganessian worked hard to keep the team's spirits up. On some nights, the laboratory staff, along with Oganessian's family and friends, would gather at his home. There, his wife Irina – a violinist who had graduated from the Moscow Conservatory – would play solo recitals. Despite the shortages and worries, her music helped soothe troubles and let spirits soar. The Americans, coming from Livermore, appreciated how hard times were for their new collaborators. When they asked if there was anything their friends wanted brought from the US, the reply was always the same: seeds. Part of the laboratory's grounds had been surrendered to the earth, claimed as a vegetable patch so the staff and their families could eat. JINR was on the brink of ruin, a hair's breadth from joining the rusted hulks of the Russian Navy, or the broken and shattered statues of Marxist heroes left on the banks of the Moskva River.

Oganessian was used to solving the impossible. In the 1970s, when the U400 cyclotron – the successor to the decommissioned U300 – had been built, he'd had to come up with a way to construct a 2,100t magnet on site. The iron, delivered from the Krivoi Rog Metallurgical Works, had come in 15m (50ft) sheets. There were no workbenches large enough in Dubna to handle the strips of rolled metal, so Oganessian had arranged for a system of rails and pulleys, moving the tooling machine over the sheet instead of the other way around. The accelerator hall hadn't been designed for something of U400's size, and Oganessian had needed to call back to his love of architecture to make everything fit. When it didn't, the team had simply smashed through the wall, jackhammering a mess of cables through solid concrete. 'The Russians,' one of the US chemists joked, 'are masters of sufficiency.'

The most essential piece of equipment for any modern element hunter was a separator capable of stripping out noise and allowing them to detect the ever-lower cross sections. In 1989 Oganessian had overseen a new gas-filled device 1,000 times more sensitive than anything the Russians had used previously: their cross section limits for detecting new elements were now around 10 nanobarns (10^{-36} m^2). The U400 cyclotron had been improved at the same time and could fire its ions with a beam intensity unmatched anywhere in the world.

With the help of Livermore, the Dubna team were slowing getting ready to go element hunting again. This time, they would use a new technique: hot fusion. This relied on the same principles as the earlier light-ion-induced reactions, but with a beam of the doubly magic, neutron-rich calcium-48. All previous calcium-48 experiments had failed because the technology just wasn't advanced enough. But by the 1990s the collaboration was confident it could be used to push the known elements onto the fringes of the island of stability.

When Oganessian gathered his scientific and technical staff together, everyone believed it was the end. There was no use for superheavy elements, and people across Russia were queuing for bread. The driving forces of element discovery during the twentieth century – the nuclear bomb, the race to harness the power of the atom and eventually the jingoistic battles of national pride – had all vanished. Science for the sake of science doesn't pay bills.

But Oganessian refused to surrender. 'We could weep,' he told the assembled scientists. 'We could shed tears. We could find excuses for our own inactivity. Instead, we are going to seek a way out of this difficult situation. We will find new sources of financing and new ways of solving arising problems.' Already he had written to foreign institutions interested in forging ties with Dubna and private sector companies eager to borrow the Russian accelerator. The new approach would guarantee funds and allow them to continue.

For the rest of the decade, the Russians balanced the need for capital with the pursuit of their dream. First, they explored and probed the regions around element 108 – hassium – to see why it was more stable than its neighbours. The answer seemed to be a smaller island of stability (rock of stability?) due to nuclear deformation – a perfect test to hone their equipment. Ken Hulet had retired, and the team from Livermore consisted of Ken Moody, Ron Lougheed, John Wild and Nancy Stoyer. Later, they brought in Nancy's husband, physicist Mark Stoyer, as an extra body. 'It took eight years to improve the apparatus, the beam intensity,' Mark Stoyer recalls. 'There was a lot that needed to be improved, and while we were doing that we could do a lot of experiments on 108.' Together, the Dubna–Livermore collaboration discovered a host of new isotopes.

At first, the team focused its search on the next element in sequence, 110. But by 1998 the team had the equipment and beam to leapfrog even further into the unknown: calcium-48 fired into a plutonium-244 target. Despite Oganessian's optimism, the team never really believed they would find a new element, Mark Stoyer says. 'We were expecting … OK, we've set limits and improved the lowest cross section [we can detect something at], but we might not see something. We were just hoping to set the world's lowest [cross section] limit.' Calcium-48 was, however, an ideal hot fusion beam, with its two magic numbers creating wonderful, stable nuclei. With the combined experience of the Russians and Americans, it finally paid off: they produced the single atom of element 114, as Ghiorso had whispered to Seaborg on his deathbed. It wasn't enough to claim the discovery of a new element – it could have been a random event, a ghost in the machine – but finally the element hunters had a glimpse of the element they had been trying to find for 30 years.

Sadly, the team did not discover the island of stability. While calcium-48 was neutron-rich, the team were only producing element 114 with 176 neutrons: still 8 shy of the magic 184 neutrons needed to hit the island. Even so, Oganessian wasn't disheartened. 'If there was no stability,' he says, '[the element 114 atom] would have had a half-life of 10^{-19} seconds. The effect we saw was a decay we could measure in seconds, 19 orders of magnitude higher.' The island was real yet remained tantalisingly just out of reach. The results were enough to convince the Dubna–Livermore team they had to continue.

The JINR record states that 'it was thanks to Yuri Oganessian that the Flerov Laboratory survived and demonstrated its vitality and resilience.' It is an understatement. Thanks to the Armenian's gift for bringing people together, the Russian programme had gone from the brink of destruction to leading the element charge into the twenty-first century. First, they would look for elements 114 and 116 before filling out the table.

The future belonged to Dubna. But in the meantime, Sigurd Hofmann's group at GSI were still leading the charge – and had already laid claim to a trio of new discoveries.

⋆⋆⋆

Germany in the 1990s was, in many respects, the reverse of Russia. Since the fall of the Berlin Wall, the two halves of a split nation had reformed, reforged and come together to become the dominant player in Europe. Reunification had been a shock to the economy and had sent it into a depression, but by 1994 things had begun to turn around and the country was buoyed with hope and optimism.

At GSI, Sigurd Hofmann's team had everything ready for another element discovery spree. The problem was beam time. The cross sections for the superheavies discovered by cold fusion were falling by around a factor of four per element. That suggested element 110 – the next in sequence – would have a fusion cross section of 1.5 picobarns. Half-lives were shrinking too. The first isotope of meitnerium found, meitnerium-266, had a half-live of just 1.7 milliseconds before it decayed. The Germans were neck-deep in the 'sea of instability', completely off the peninsula where nuclides were stable. To produce a single atom of meitnerium, GSI needed to run their beam continuously for two weeks. Under the same conditions, finding element 110 would mean continuous running of the machine for almost six months without breaks – 24 hours a day, every day.

Such a challenge would have been daunting even if the Germans could commit to it. But GSI's accelerator was in demand for other projects – there was more important research to do than fire ions at a lump of lead and hope something would happen. No one doubted that element 110 was there to be found. The argument was that it had no practical use and the experiment had no known purpose; even if the Germans did find it, all they would be able to do is say 'hey, look what we did'. In some eyes, it was a vanity project that would cost millions.

The only way to get the project greenlit was to make it more efficient. 'We started thinking about looking for 110 in 1988,' Hofmann recalls. 'It was my work now. We had to gain a factor of 10; we had to reduce the beam time from 150 days to 15 days.'

It took five years to improve the machine. 'Behind where the beam stopped, we had water cooling,' Hofmann says, remembering the result of one of their upgrades. 'When we fired the beam, one of the plates failed. The beam intensity had become so strong that we immediately burned a hole into it.' Perhaps the smartest innovation was to add a deliberate bend in their detector, SHIP. After passing through electrical and magnetic deflectors, any created elements would have to navigate a seven-degree kink in the pipe, aided by deflector magnets (anything else would end up slamming into the wall). It reduced the background noise, the fission debris and flying ions, by 90 per cent.

Hofmann's team had also picked up the new darling of the superheavy world. Victor Ninov was a whiz-kid from Bulgaria who brought a raw, infectious enthusiasm to the team. Young, energetic and with a short crop of curling black hair, the researcher was a Renaissance man: gifted at whatever task he turned his hand toward, whether it was science, music or sport. Ninov had a wacky sense of humour – his standard sign-off on emails was 'Your crazy Bulgarian'. His hobbies were equally eclectic. Along with Matti Leino (now at GSI), he had taken to 'testing' the Italian restaurants in Darmstadt, ordering the same plate of spaghetti carbonara in a quest to find the best pasta in town. 'We were quite close,' Leino says. 'He was supposed to come and play violin at my wedding but had a bicycle accident and hurt his violin thumb.' Another passion was mountain climbing, and Ninov would stay at Heinz Gäggeler's home in the High Alps of Switzerland between experiments. 'He was brilliant,' Gäggeler remembers. 'He was the enthusiastic rising star … we all liked him.' At work, Ninov's role in the team was critical. 'He was quite the expert in programming computers,' Hofmann says. 'In 1988 we got new computers and Victor became *the* expert.' Ninov devised the programme, 'Goosy', that would analyse the 'hits' and report any new elements electronically. Gone were the days of racing a Volkswagen Beetle to see what you had made or crowding around analogue monitors hoping for a ping.

Toward the end of 1994 a slot of beam time with the GSI UNILAC opened. Hofmann was faced with a tough decision: how hard should he fire the accelerator? For the past two decades, GSI's team had been split about what would happen with the nucleus of superheavy elements. It was clear the finely balanced forces of repulsion and attraction that kept a nucleus together had started to shift as the elements got heavier. That gave the team two options. Either they could increase the energy of their beam – a theory called 'extra push' – or lower the energy to sneak over the Coulomb barrier. 'Armbruster wanted to increase the energy [of the beam],' Hofmann remembers. 'I wanted to decrease it. We just couldn't decide.' Getting it wrong at this point would have meant the whole project would have been a waste: they'd spend their time searching in the wrong direction.

The GSI team decided to experiment. 'We needed to measure the excitation functions very accurately. We decided to produce hassium, element 108, before we did 110. We needed iron and lead,' Hofmann recalls. 'The problem is that we needed 4 grams of iron-58 if we were going to run for three or four weeks. At that time, 1 gram cost 500,000 Deutschmarks.' Today that's worth around £500,000. The Russians came to the rescue. 'Dubna sent us the iron, almost 21 grams of enriched iron. I kept it in my desk – I was the richest man in Darmstadt!' The Germans, aware that the Russians were struggling, didn't hesitate to return the favour, sending Oganessian's team electronics and detectors that they just couldn't get elsewhere. Oganessian's new approach of joint working was bearing fruit.

The experiments with hassium helped, but it still wasn't clear what the optimum beam energy would be. Meanwhile, a queue was building up at GSI: another team needed the accelerator. Hofmann had a window of four weeks to find the elusive element 110 but was still debating the best way to go with Armbruster. In the end, Hofmann decided to take matters into his own hands. 'Armbruster had been away in Grenoble, and I'd made all these improvements without him. So, I thought I'd do the experiment without him too!'

On 9 November 1994 Hofmann's team changed the iron to nickel and lowered the intensity. 'After one day, we had our first decay chain. One day! Four weeks later, we had the whole decay chain.'

Once again, after a gap of 10 years, elements were being uncovered at GSI. Hofmann, Ninov and the rest of the team were delirious.

Then they heard even more exciting news: the experiment that was queued up to use the accelerator after them wasn't ready. GSI's superheavy element hunters had another 17 days of beam time. Having discovered one element, why not make it two? 'This was at the end of November,' Hofmann says. 'I thought I'd put in a bismuth target [one proton heavier than lead] to make element 111. By the end of December, we had three decay chains for that too. It was a good year.' The news came not a moment too soon: on New Year's Day 1995 Münzenberg received a phone call from Yuri Oganessian. The Russians had also been looking for element 110; GSI had been barely a month ahead of the Dubna–Livermore team.*

Hofmann's two new elements, 110 and 111, were undisputed. Now, the team – physically and mentally drained after 24-hour stints at the beam line – wondered if they could push any further. The cross sections had continued to plummet, and the half-lives of the newly created isotopes had slipped to as low as 170 microseconds. It was unlikely they would succeed, but in the end the temptation to at least try for another element was too strong to pass up. In January 1996 Hofmann, Ninov and the others were ready to go again,

---

* Berkeley also had an (albeit weak) claim for element 110; using cold fusion, Ghiorso *et al.* reported making a single atom in June 1994. But, as Ghiorso himself would concede, one atom was not enough proof, and in element discovery there is no second place.

this time firing zinc into lead to make 112. To the shock of everyone, GSI seemed to strike gold almost immediately.

'After one week, Ninov came to me and said we'd observed something,' Hofmann recalls. 'I told him "OK, let's have a look," and asked him to print out the raw data. Energy, time, position. It was relatively simple to make such a printout. It was about lunchtime, and Ninov said, "Yeah, I'll do it after lunch." He didn't do it, so I asked him again – it was just one command on the computer – but he just said "Yeah, yeah, I have no time now." He came several hours later with the printout. Some of the data was missing and it wasn't what I expected for a decay chain. I told him we couldn't publish it, we'd have to wait for another event.

'A week later, we got a perfect decay chain for element 112. Everything agreed: the energies and the positions. We were really happy. This was our main publication; the earlier [Ninov] chain was simply mentioned in the paper.'

There is an uncomfortable pause.

'It is fortunate we did it this way.'

CHAPTER SEVENTEEN

The Ninov Fraud

Darleane Hoffman sat in her office on the hilly rise above the Berkeley campus, admiring the view across San Francisco Bay. In the distance she could look out and see the old federal prison on Alcatraz, the city's main tourist attraction, and past that, the rolling fog slipping like white sheets under the Golden Gate Bridge. It was Monday, 19 April 1999. It would have been Glenn Seaborg's eighty-seventh birthday.

Almost 50 years earlier, she had missed the discoveries of einsteinium and fermium while she sat outside Los Alamos, waiting for HR to realise that women could be scientists too. Now, her team leaders wanted to tell her something important. She had feared the worst but had been reassured on the phone that it was good news.

In her early seventies, Hoffman gave the impression of being someone's sweet grandmother. Foolishly, some graduates tried to get into her group assuming she'd be a pushover. It was an opinion that was quickly dispelled – particularly if you dared to call element 105, officially dubnium, anything other than 'hahnium'. Tough and supportive in equal measure, universally respected across the chemical world, Hoffman had dedicated her life to making sure nuclear chemistry wouldn't die, even if she had to resuscitate it one student at a time.

Berkeley was still playing catch-up with the other labs, but finally seemed to have turned a corner. Super-HILAC had been turned off in 1993, and the experiments had moved to the 88-inch cyclotron; SASSY had also been replaced by a new gas-filled separator (this time with a more fitting emergency valve). With it, the team hoped to hit even lower detection limits.

Matti Nurmia had returned to Finland. In his place came three additions. The first was Darleane Hoffman's former postdoc Ken Gregorich. Tall and wiry, with a neatly trimmed goatee and a crop of hair skirting a bald crown, Gregorich was the epitome of the 'work hard, play hard' attitude of nearby Silicon Valley. In the lab he was a relentless and meticulous researcher; at home he ran ultra-marathons for fun. He had worked with Hoffman since the mid-1980s and was at the vanguard of a new generation not poisoned by the Cold War. Sharp, measured and precise in his approach, to Gregorich the transfermium storms were water under the bridge. He just wanted to do good science.

Next, on a Fulbright scholarship from the Soltan Institute for Nuclear Studies in Warsaw, Poland, was Robert Smolańczuk (Berkeley didn't have the budget for another permanent member of the team, so the secondment was the best solution). A theoretical physicist, while Smolańczuk had been at GSI he had published some eyebrow-raising calculations. According to his theory, the cross sections for the elements beyond 114 wouldn't vanish into the realms of statistical improbability, but rather be large enough to detect: a massive 670 picobarns. In the 1940s such a low limit would have felt impossible; by the turn of the millennium, the element hunter's toolbox had improved so dramatically it felt like it was easy pickings – the kind of cross section that could produce hundreds of atoms a week. If Berkeley was willing to give Smolańczuk a shot, he was confident they could leap beyond anything Dubna were trying to do and find element 118. It was controversial, flying against all known wisdom. But then, wasn't that what element hunting was all about?

The final new arrival was considered a coup: Berkeley had tempted element superstar Victor Ninov to leave GSI and join their team. Ghiorso, technically retired but still doggedly riding his recumbent bike into work every morning, saw the newcomer as the future of element discovery. 'Victor Ninov,' Ghiorso would tell anyone who would listen, 'reminds me of a young … well, a young Al Ghiorso!' On the back of her colleague's glowing recommendation – and letters praising his talent from his colleagues at GSI – Hoffman had placed complete faith in him; while Gregorich led the team in running the machines, Ninov had brought over his unique computer program from GSI to analyse the results. He was the only one who knew how it worked, but Berkeley didn't need anyone else: he was the best in the world at what he did.

Hoffman's hand had been twisted toward Smolańczuk's madcap idea. When Dubna claimed element 114 had been found, the Berkeley team had been a mere eight months behind them, ready to do the same experiment. However, they hadn't been able to get hold of the large quantities of plutonium-244 or calcium-48 needed (or the permission required to use plutonium in the hills above one of the most populous metro areas in the US). Options narrowed; all that was left was 'Robert's reaction': firing krypton into lead.

Hoffman and Ghiorso had both urged doing it as soon as possible – if Smolańczuk was right, there was nothing to stop GSI or Dubna doing it first. '[It was] a strange reaction that no one thought would go,' Ghiorso would later recall for the *New York Times*, 'but because it was relatively easy, we thought, "What the heck, we have nothing to lose."' Gregorich had agreed, arguing that the efficiency of the 88-inch cyclotron was so high that even if they didn't find 118, it would give him a chance to improve its systems. Finally, Ninov relented and threw himself into the analysis with his usual verve.

The experiment had started on 8 April 1999 and ran for four days. At first, nothing happened. The team departed for the Easter break, leaving Ninov to check their results. Now, almost two weeks later, Hoffman watched as a trio of researchers – Gregorich, Ninov and Walter Loveland, on sabbatical from Oregon State University and there to soak up how Berkeley conducted their experiments – entered her office. They brought with them a sheet of data.

While the experiment ran, Ninov's analysis tool, Goosy, had found three distinct alpha decay chains resulting from fusion. Two of them matched Smolańczuk's predictions perfectly. The numbers coming from Ninov's analysis were too good to put down to random chance.

Ninov laughed. 'Does Robert talk to God, or what?'

Berkeley had discovered element 118.

The reaction from the team had been a mix of excitement and disbelief, even before their results had made their way to Hoffman's desk. Loveland's first response had been 'What the hell is going on?'; Gregorich was surprised too; Ninov had been so taken aback he had urged his collaborators to keep the results quiet and not tell Hoffman (Loveland and Gregorich overruled him). For her part, Hoffman felt a pinch of excitement, but kept her cool. Science, she knew, thrives on verification: the experiment had to be repeated or the results were meaningless. She was too seasoned to get her hopes up on what could be a phantom.

'OK,' she said. 'Let's do it again.'

The Berkeley team started a second experiment running 'Robert's reaction'. By the first week of May, they had another perfect chain that matched Smolańczuk's calculations (the previous chain that didn't follow the pattern was discarded). Dubna's single atom of 114 wasn't enough to convince IUPAC of a discovery; Berkeley's three atoms were solid, unchallengeable proof. Hoffman and Ghiorso began to dream of snatching another element out from under Oganessian's nose. Already, thoughts turned to its name. Berkeley had seaborgium; why not 'ghiorsium' too?

Wary of the false reports of discoveries throughout the Cold War, Berkeley Lab decided to proceed with caution, conducting an internal review to rule out any embarrassing mistakes. The staff double-checked everything: the ion source, the accelerator and the detectors. Nothing was wrong. Finally convinced, in June 1999 Hoffman and Ghiorso called a press conference and published their claim in full. Anyone who had been near to the experiment was added to the paper, with Ninov as first author. 'Needless to say,' Hoffman and Ghiorso wrote in *The Transuranium People*, published that year, 'this news is an enormous surprise to the scientific world. Now there is no question, the Superheavy Island [of Stability] actually exists! […] We have convincing evidence of 114, 116 and 118! This opens up a whole new region for study.'

Finally, on Glenn Seaborg's birthday, Darleane Hoffman had her element. It seemed too good to be true.

It was.

Hoffman, Ghiorso and their team had just fallen victim to the most audacious fraud in science history.

★ ★ ★

Every scientist makes mistakes. Science functions by 'failing forward', constantly tinkering ideas, experiments and approaches to get things a little closer to right each time. Research fraud – faking your results, lying to yourself and the world – is the opposite of everything science stands for. When you're caught (and you always are), it destroys your reputation, your colleagues' reputation and your lab's reputation. In physics, the best-known scandal is probably the work of Jan Hendrik Schön, who seemed to have made miracle breakthroughs in semiconductors that were nothing of the sort. When news of his scientific misconduct broke, it saw 28 papers in leading journals retracted.

The 118 scandal, which occurred at virtually the same time, was more devastating for its community. The first cracks in the Berkeley discovery had emerged within months of the paper's publication. Back in Germany the GSI team were struggling to repeat the Berkeley results. The institute's superheavy programme had not had a success since 1996 and the new director, Hans Specht, had quarrelled with Sigurd Hofmann. 'He gave us a hard time,' Hofmann recalls. 'We immediately got the beam time to repeat the experiment for 118; krypton is an easy beam [to produce], and lead targets are easy too. After one week, when he hadn't seen anything, our director shouted at us: "You're not able to make such experiments, you're too stupid!" But after another two weeks, we still didn't see anything either.' (Specht told me he does not remember this, although it was clear when we corresponded there was no love lost between him and Hofmann.)

Teams in France and Japan then tried the experiment. Neither found any of the miraculous decay chains the

Americans had reported. Stranger still, when Ninov attended conferences, he was reluctant to speak about the amazing feat Berkeley had accomplished. Questions from the audience were deflected, sidetracked or, as had happened at GSI, casually forgotten about. The man of the hour seemed to want to avoid all mention of his greatest achievement.

In the spring of 2000 the Berkeley team decided to silence their critics and run the experiment again. What had once been so easy to produce became invisible: the 118 decay chains weren't there. For the next year, the laboratory went over the data, even calling in an independent group to analyse the experiments and make recommendations.

In April 2001 the team gave their amazing krypton-into-lead reaction another shot. Two-thirds of the way through their beam time, Ninov gave everyone the news they had been waiting for: his analysis showed clear evidence of a decay chain for element 118.

The news should have brought relief; instead, it brought the opposite. 'In the time that passed,' Loveland recalls, 'there were other people who had become expert in Goosy, including my postdoc, Don Peterson.' Keen to make sure everything was correct, Peterson decided to go back into the raw data, before it had passed through Ninov for analysis, to check the chain was accurate. 'Don looked at this stuff and said, "I can't find the event!" At that point, I thought "Oh God, what's happening?"'

Loveland checked over his colleague's work, pulling up the original data from the machine itself. Peterson was right: the alpha decay chain Ninov claimed to have found wasn't there. 'Depending on whose software you used, if Ninov used it or Don Peterson used it, you got different answers,' Loveland explains. 'And that is just *not right*. At that point I started yelling to everyone that something was terribly, terribly wrong.'

In June 2001 the team went back to review the original 1999 data tapes – the raw, unfiltered information spewed out before it had been processed by Ninov. None of the chains he had taken up to Hoffman's office existed. A third Berkeley committee agreed: there was no sign of element 118. There never had been.

For decades, Berkeley had openly mocked and questioned the Russian data. Now their own research was flat-out wrong. What had been the highlight of Hoffman's career and the crowning glory of Ghiorso's had become the darkest moment of their lives. Meeting with the group, everyone except Ninov agreed to retract the claim. While Ninov conceded the records didn't show his element, he remained adamant about what he had seen during his analysis.

Nobody was listening to him. Goosy had a habit of spawning quirky, corrupted data, but a review showed it hadn't made a mistake; and, even if it had, the probability of spawning three chains that fit the predictions *exactly* wasn't worth contemplating. The different committees also ruled out the possibility that someone had wiped the event data from the original tapes. That left only one possibility, the final committee decided:

> *There is clear evidence that at least one of the 118 element decay chains published in 1999, and also the candidate in the 2001 data, were fabricated. This fabrication was performed by capturing the output of the data analysis program in a text editor and then systematically altering some events and inventing others in order to present data that would appear to be an element 118 decay chain.*

Someone had, in 1999 and 2001, copy-pasted evidence of element 118 into the raw data. Someone analysing data only one man saw, and who knew how to use a computer program only one man could interpret.

'In short,' the committee found, 'it is very difficult to reconcile all these circumstances on any basis other than with Ninov being the fabricator of the claimed 118 decay chains.' In 2002 Berkeley Lab found Victor Ninov guilty of scientific misconduct and fired him.

The recriminations were swift. The rest of the team received harsh criticism that only one person had been left checking the results, creating a weak link where someone could interfere with the scientific method. 'I had hired a world-recognised expert and we were trusting him to do a job,' Gregorich told the *New York Times*. There were no safeguards because what happened was so unthinkable.

Ghiorso's appraisal was blunt. 'It's good that Seaborg died before this,' he was quoted as saying in the same article. 'He would have been one of the co-authors. This would have just about killed him.'

* * *

At GSI, elements 110, 111 and 112 had all been confirmed after Ninov had departed, removing any possibility that the German claims were also faked. Even so, Hofmann wanted to make sure everything was correct. He asked a colleague to go back through the old files from the 'miracle years' of 1994 to 1996. Had anyone faked the data in them too?

'It took about three or four hours,' Hofmann told me. 'We looked, and what existed was an alpha decay of polonium into lead.' This was chicken feed for the element team: just the random radioactive noise you'd expect in a particle accelerator. 'Then we looked at the subsequent printouts. On the old computers, they had version numbers, and we found them with the help of our computer centre. There were different versions in Ninov's old computer ... he had changed the information from the background event into 112. He had added, step by step, additional numbers to make it look like an alpha decay chain from 112. It wasn't completely accurate, and I could see something was wrong.'

Hofmann remembered the events of 1996, when Ninov had burst into his office before lunch with the 'discovery' of element 112, only to hesitate for hours before giving him a simple printout. The strange delay suddenly made sense: had Ninov rushed off to fabricate the data?

'I told the new director [Walter Henning, who had replaced Specht],' Hoffmann explains. 'He gave me some good advice:

look through *all* the data Ninov was involved with. And we found a second chain, from the discovery of 110, that had been manipulated too.' Of 34 chains, the two fakes stood out like sore thumbs. GSI quickly published a retraction of the altered chains, euphemistically referring to 'inconsistencies' in the data. Everyone knew exactly what they meant.

Fortunately for Hofmann and the Germans, they had always treated Ninov's spurious chains with caution, and the rest of their work was beyond reproach: the two bad apples were removed before the whole barrel was tainted. Within a decade, the GSI elements were all confirmed as real.

The GSI elements had names too. On 10 December 1997 Hofmann declared a 'names-finding day' at GSI, gathering up his team, as well as Matti Leino and guests from Dubna and Bratislava. Hofmann went through the names sent in by colleagues as suggestions, with Andrey Popeko writing them down on the blackboard. The team then went around the table, finishing with a list of 30 possibilities. 'The arguments raged fast and furiously,' Hofmann wrote. 'Painfully, one at a time, names were ringed with red chalk until, in the end, we had found the three we needed.'

Element 110 became 'darmstadtium' (one suggestion that was rejected, which came from an American school class, was 'policium' – in Germany, 110 is the emergency dial code for the police). For 111 and 112, GSI chose to honour famous scientists throughout the ages. Element 111 would be 'roentgenium', after Wilhelm Conrad Röntgen, the discoverer of the X-ray; and element 112 would be 'copernicium', after Nicolaus Copernicus, the Renaissance scientist who showed that the Earth orbited around the Sun.

They would be – to date – the last elements discovered at GSI. In 1999 the Germans had dreamed of creating a 'superheavy element factory', possibly exploring as far as element 126. Hofmann asked for permission to start using calcium-48 and actinide targets too, confident that GSI could easily overhaul the Dubna–Livermore team. His request was turned down. 'We could have started experiments in hot fusion,' Hofmann told me. 'But the proposition was rejected, and not with friendly words. It was a dead thing.'*

* * *

Victor Ninov has always insisted on his innocence. His response to the Berkeley investigation in February 2002 was frank: 'At no time did I knowingly engage in any form of misinterpretation of data or scientific misconduct [...] I have never intentionally altered, invented, fabricated, corrupted, deleted or concealed data [...] I stand by the integrity of my research.'

Having criss-crossed the world, I have never met a heavy element researcher who believes him. Most just want closure. Most, simply, want to know why.

It's a hard question to answer. The glory of discovering an element doesn't make a very good motive; the chance of a made-up alpha decay chain perfectly fitting the real results (which would have emerged eventually) and fooling a seasoned element hunter was virtually zero; Sigurd Hofmann's immediate, at-a-glance dismissal of the fictional chain for element 112 was proof of that.

Nor was there pressure to succeed. No one, at any point, suggested Ninov's career was in the balance if another element wasn't found. He had already discovered three elements. He had nothing to prove.

Al Ghiorso, speculating, suggested that Ninov's intent had been to buy time for Berkeley: by inserting the false chain, it won the team more beam time to find the *actual* element 118. But again, the argument doesn't hold water. Robert Smolańczuk's ideas were considered a Hail Mary – nobody had really believed they would succeed. Why buy time for an experiment that none of the team thought would work?

---

* I've seen the internal report, and Hofmann isn't kidding. However, it would be misleading and disrespectful to everyone involved to say this was the only reason GSI fell behind in the element race. The truth is more complicated.

'None of us fully understand what Victor did or why he did it,' Loveland says. 'He was extremely well thought of, a very talented man. If you ask me why, I have no real idea. Perhaps he became overconfident in his ability to predict what was going on. I was never able to come to an understanding with him. I spoke to him on a daily basis. At one time he alluded to other people interfering instead of him, but that was nonsense.'

The best guess comes from Sigurd Hofmann. 'The astonishing thing is that [the first time data was fabricated, for element 110] it happened on 11 November. The chain Victor produced had a half-life of 11.19 minutes. If you know something about German carnivals, you'll know they start on 11 November at 11.11 a.m. I think he meant it as a joke. But, with such a joke, he realised he could manipulate things and nobody would realise it.' Perhaps it isn't such a surprise element 118 was 'discovered' on Glenn Seaborg's birthday after all.

For Loveland, the Ninov episode emphasises a positive: 'Science works. You get an anomalous result, and if it's right it gets stronger and stronger [as the experiment is repeated]. Sometimes it doesn't. There have been other cases where outstanding events were announced and pulled back because they couldn't be reproduced. In this case, it has the additional element that there appears to be fraud involved.' In the long term, science self-corrects, finds answers and pushes forward. Ninov was caught and the discovery was retracted. Time to move on.

Even so, the Ninov scandal sent shockwaves through the superheavy community. Heinz Gäggeler, the man he had once called a friend and who had put Ninov up in his house while climbing the High Alps, feels betrayed. 'Victor was so well received when he came to Berkeley,' Gäggeler told me. 'He had full support. And because of that, one didn't look too carefully into the analysis he was doing. It was a total disaster. Did it destroy Berkeley? Of course it did. Berkeley was *Berkeley*. The outside world doesn't want fake news. The show was over.'

The scandal brought an ignominious close to Al Ghiorso's remarkable element-hunting career. For Ken Gregorich, the fallout was even more devastating. Today, 20 years down the line, it remains a sore subject on what has otherwise been a glittering career. When I asked him about it, he politely declined to go over it again. 'It was a dark period, and it's gone, and I'd rather leave it at that.' I can't blame him.

Yet greatest in the team's sympathies was Darleane Hoffman. She had been denied part of the discovery of einsteinium and fermium because of sexism; now, so close, her dream of being an element discoverer was over. 'When we thought we had [element 118],' one former Berkeley researcher told me, 'we had it for her. It was Darleane's element. And that crushed us more than the retraction – that she didn't have it.'

CHAPTER EIGHTEEN

A New Hope

In 1977 Dawn Shaughnessy saw the greatest thing ever. Like most young girls living in California in the late 1970s, her friends were obsessed with Barbie dolls. Shaughnessy's passion was *Star Wars*. The first time she read those yellow words as they scrolled up the screen, heard the John Williams score kick in and witnessed a Star Destroyer chase down a fleeing Princess Leia, she was blown away by the spectacle of it all. The movie had everything: lightsabers and space battles; dusty deserts and cold space stations; wise words from Obi-Wan Kenobi and the staccato menace of Darth Vader. She watched the film 20 times in the cinema that year, then wrote away for the action figures as they trickled into production throughout 1978. She ripped open the packets and started re-enacting adventures from a galaxy far, far away on the floor of her bedroom. 'Forget Barbie,' she told her friends, 'this is *way cooler*.'

In 1985 her family moved to El Segundo, nestled between the Santa Monica Bay and the fringes of Los Angeles International Airport. By then, the *Star Wars* craze had passed, the neighbourhood kids having moved on to the next cinematic spectacle or transforming robot. Shaughnessy stayed true. She'd seen the Galactic Empire blast Alderaan to pieces and wanted her chance to play with big freakin' lasers too. That meant becoming a scientist.

Like thousands of youngsters up and down the US, Shaughnessy received a host of science kits from her parents – everything from simple circuit boards to small-scale laboratories. Most children who receive such gifts forget about them after a few days of goofing around, or end up grounded for blowing fuses and accidentally spilling iron filings over the carpet. Shaughnessy took the kits, finished them and wanted more. El Segundo High's chemistry lab was threadbare, its science classes lacklustre, so she began to make her own experiments at home. Eventually, she made it to the undergraduate chemistry programme at Berkeley. There, she fell in love with nuclear science and wanted to be part of it.

She was a decade too late: in the early 1990s nuclear chemistry was in decline. As mentioned earlier, Three Mile Island and Chernobyl had sapped political will, while the secretive nature of the work had suffocated and smothered publication to the point of excluding outsiders. The great funding boom had vanished, while the national labs had turned their focus to other projects. 'Oh great,' she told herself, 'I've picked a really stupid area.' In frustration, she went to talk to one of the few established scientists left standing and asked for help.

The professor she approached was Darleane Hoffman. In Shaughnessy, the elder stateswoman of the superheavy world saw the kind of researcher the community needed: talented, focused and driven, with the same desire to have a normal life outside the lab as she had fought to keep almost half a century earlier. 'Come here,' Hoffman told her, 'and go to grad school. It'll be totally different. You'll be on the hill with me.'

It was an interesting education. Shaughnessy's speciality was the environment; the US had realised that it hadn't put together a proper plan to contain the radioactive waste from its facilities, and the focus of research had shifted toward clean-up operations. Yet under Hoffman, Shaughnessy had even more exciting opportunities. Eventually she became involved in the Berkeley superheavy element hunt, pulling late-night shifts on the 88-inch cyclotron while Ken Gregorich and Victor Ninov made their adjustments. Superheavy science was almost like modern alchemy, but then again, hokey religions and ancient weapons are no match for a good blaster at your side. At the end of the day, wasn't that all the 88-inch cyclotron was?

Then came the announcement. Element 118. Shaughnessy and the gathered students and postdocs were ecstatic. Still in her twenties, still completing her PhD, Shaughnessy had made it onto a paper that would be remembered for all time.

Remembered, it soon emerged, for all the wrong reasons. As the Ninov scandal shook the Berkeley Hills, Shaughnessy – like every other innocent researcher caught up in the affair – found herself fighting to save her scientific career.

★ ★ ★

'Yeah … try looking for jobs when you have *that* on your CV.' Shaughnessy rolls her eyes as she remembers the fallout of the 118 retraction. 'The community is so small that when you went around talking to people they'd say, "Oh, you're from *Berkeley* …" My first meeting here at Livermore, Nancy Stoyer said: "You must have been on the Ninov paper." I was like, "Yeah, thanks for bringing that up …"'

Now in her forties, Shaughnessy cuts a relaxed figure, peppering our chat with quips and the sheer, infectious joy she takes from her research. Her facial expressions switch from seriousness to puckish charm. She wears the loyalties ingrained since childhood – science and *Star Wars* – on her sleeve. When *The Force Awakens* came out, her team had a contest to see who'd watch it the most at the cinema. Shaughnessy won by double digits. My geek test was whether I knew who Ahsoka Tano was. My pass for the lab (along with the kind of security clearance you'd expect for a place still at the heart of maintaining the US nuclear deterrent) was a pledge to send her a poster signed by Darth Vader actor David Prowse.

Livermore is an hour and a half on the Bay Area's metro system from Berkeley. It swings you through downtown Oakland, past the baseball stadium and along the waterfront, before a change in trains sneaks you under the ridges that enclose the Bay and out into the brilliant sunlight of the Livermore Valley. It feels like stepping into a Californian version of Narnia: in an instant, the cloudy cityscapes, chill ocean air and huddled commuters strip away into bucolic peace. Here, basking among gold-spun hills, the pace is relaxed and easy. The mayor, John Marchand, describes it as a city of 'contrasts and superlatives'. 'We are the oldest wine region in California – you buy auto parts at Napa, in Livermore we make wine. We have two national labs. We've had the world's fastest computer, we still have the world's fastest rodeo: a bull ride is 8 seconds, but with 22 gates they keep the thing going for 3 or 4 hours. We even have the world's longest-burning light bulb. Been burnin' [since 1901].'

He's not kidding. The bulb has a direct webcam feed: 24 hours a day, 7 days a week you can tune in to see if it's still glowing. But the real prize is Lawrence Livermore National Laboratory.

When it was built in 1952, Livermore was merely an offshoot of the Berkeley lab. By the end of the decade, it was developing the Polaris intercontinental ballistic missile. Today it's the heart of America's research efforts against bioterrorism, nuclear non-proliferation, energy and environmental security. On 2.5km^2 (1 mile^2) of campus, 5,800 staff work with a budget of $1.5 billion. In the nuclear science programme, 235 scientists work with $90 million – 80 per cent tied directly to real-world applications, 20 per cent tied to more fundamental research. Livermore honours Ernest Lawrence's dream of Big Science in action.

None of its staff get to work on superheavy elements completely: it's a side project to keep them excited. Shaughnessy spends part of her time at Livermore's National Ignition Facility. The largest, most powerful laser ever designed, it is the size of three football fields, using nearly 40,000 optics to guide 192 laser beams to focus on a target the size of a cotton bud. The result is a beam of more than 100 million °C, or pressures more than 100 billion times the Earth's atmosphere. When it fires, it uses more energy than the rest of the entire United States. It's the closest thing Earth has to the Death Star. Small wonder Shaughnessy is all over it.

Shaughnessy completed her PhD at Berkeley in 2000, then moved over to Livermore two years later as a postdoc, where she joined the Dubna–Livermore collaboration.* By that time, the team had come up with the perfect riposte to anyone who suggested that the Ninov scandal could spiral out and affect more than just Berkeley and GSI. 'We went about the data upside down,' Shaughnessy explains. '[The Russians] looked for alpha that ended in fission; we looked for fissions and then went looking for alphas. And if you get the same chain, it's much more convincing.' At Livermore, chemist and physicist duo Nancy and Mark Stoyer created a Monte Carlo simulation. Taking its name from the Monte Carlo casino (where, according to legend, the uncle of the simulation's creator used to borrow money from relatives to gamble), the idea was to use randomness against itself. By running hundreds, thousands, millions of scenarios in a computer and aggregating the results, you can quickly find out what you're looking for. Meanwhile, in Russia, scientists adopted the exact opposite approach and scoured the data for signs of actual events. The results weren't just being checked by two different people, but by two different labs using completely different methods.

As Berkeley recovered from the fallout and GSI found its programme curtailed, the Dubna–Livermore group hit success after success. With both Russian and American expertise – and the doubly magic, neutron-rich calcium-48 beam – the elements were just waiting to be found. By 2002 the team had clear evidence for fusion of elements 114, 116 and 118 (even-numbered atomic elements are easier to make).

* 'We recognised that the younger scientists on the [Ninov] paper probably would not have had much voice to object or change things,' recalls Mark Stoyer. 'The third or fourth author [on a paper] isn't mentioned much with regards to discovery of an element, so those authors shouldn't get huge amounts of blame either.'

By 2003 the team had created 115;* there was even evidence for 113, as an alpha decay product. All of the elements showed longer half-lives than would have been possible without the island of stability. But, again, none of the elements could land on it. There were too few neutrons.

With hot fusion, Dubna were once again leading the charge into the unknown. And once again, the atmosphere lingering from the transfermium wars began to rear its head. Bitter sentiments about the Cold War and mistrust began to leak out in once-civil conversation. 'When we first announced our 114 discovery there were some very complimentary comments,' recalls Nancy Stoyer. 'But once Berkeley announced their 118 discovery, all of a sudden everything became very negative. Blackball isn't quite the right word, but there was this attitude: "You're working with the Russians, you can't do anything right." Granted, the first atom we saw, the chance it was random was high. But we were honest in presenting that.'

Inevitably, tensions boiled over. In 2003, at a conference in the Napa Valley, Yuri Oganessian was giving a talk on element 114 when Hoffman rose and added her own acetate to the projector: 'None of this has been confirmed.' ('Never put that community of scientists near that much wine,' one heavy element researcher told me, only half-joking.)

Hoffman had a point. As with all element discoveries, any results would be single atoms, produced in a whirlwind of radioactive background, fissions and ions flying everywhere. Inevitably, half-lives are going to differ, isotopes are going to be misidentified and gently corrected and supposed discoveries are going to fade in favour of the right results. It's

* Element 115 has a strange claim to fame. In 1989 a UFO theorist called Bob Lazar told a Las Vegas TV show that he worked in the mysterious Area 51, where he says he reverse-engineered alien spacecraft for the US government. These UFOs, Lazar insisted, were powered by the then-undiscovered element 115. But this is a popular science book, not a conspiracy 'zine …

why science thrives on repetition, repetition, repetition. As the theoretical physicist Witold Nazarewicz put it in the *New York Times*: 'One has to be extremely careful [...] this is not because one is doing something wrong, it's because these are very difficult measurements. They are playing on the edge of statistics.'

'The rest of the community did not buy 114 at all,' Shaughnessy remembers. 'Like, forever. I thought it was never going to get confirmed. We'd done so many experiments, and repeats, and excitation functions within our group, and it was so consistent. We weren't really going "Dang it, we need to get these confirmed!" because we felt we had done real science.'

In March 2009 Livermore scientist Ken Moody attended the American Chemical Society meeting in Salt Lake City to collect the highest honour the nuclear chemistry and technology division gives out – named after (who else?) Glenn T. Seaborg. The division chair at the time was Mark Stoyer ('it's the only time Ken has ever worn a tux') and Shaughnessy had organised a programme on Moody's behalf. 'So here comes a surprise,' Shaughnessy recalls. 'Ken Gregorich came up to us, and said: "Here's some data, hot off the press, no one has seen it. We've just confirmed your discovery of element 114."'

Putting aside their differences, Berkeley had focused on the science. While they didn't have the resources or beam time to hunt for elements themselves, confirming a reaction was easier: they could calibrate their equipment to match the energy the discoverers claimed, run the reaction and see what happened. Finding a spare moment in their own research programme, Gregorich's team had run the Dubna–Livermore hot fusion reaction and come up with the same numbers. 'I fell off my chair,' Shaughnessy remembers. 'Nobody knew it was coming. You had an element where someone had confirmed someone else's data for the first time in decades.' The Cold War, the transfermium wars, were finally over for good.

But something even more important was happening at Livermore and Dubna. The superheavy scientists had worked out how to perform chemistry on their fleeting children … and it seemed like the new elements broke all the rules.

★★★

At 5 a.m. on 14 October 2016 Robert Eichler got up, donned his warmest jacket and set off into the Dubna night to fix an experiment. Tall, broad-shouldered and imposing, the giant trudged through the sub-zero climes toward the JINR gates. Eichler had virtually grown up in the city – his father was a visiting scientist from East Germany – but this was a new experience even for him.

Wearily, Eichler made his way through the armed checkpoint, up into the guts of the Flerov Laboratory of Nuclear Reactions. Up a flight of stairs, past the neat and ordered halls of Oganessian's office, into the concrete maze of the accelerator labs. Down dusty halls with cracked tiles and marked with Cyrillic warnings, past twirling alarm bulbs and humming lab set-ups, out to an experiment resting on a small metal platform – a balcony – suspended over the gloomy darkness of an interior shaft.

Eichler had travelled to Dubna from the Paul Scherrer Institut. His team of Swiss, Russian and Japanese scientists had won a month's beam time to conduct an experiment using element 114. Three hours before it was due to shut down, the system had failed. Eichler knew nothing had been lost – his team had been packing up to go home, so all their logs were safe and secure. He also knew that element 114 was produced at around a single atom a week; it was unlikely he'd get another hit anyway. But for superheavy researchers, hours, minutes and even seconds matter. Beam time is the most precious commodity, the source of all life and stress. Eichler wasn't going to let personal comfort cost him any more than necessary. Not with stakes so high.

Eichler's experiment was, for a chemist, one of the most tantalising prospects possible: he was looking for the collapse of the periodic table.

The periodic table is based on trends. If you follow the columns down, the properties of the elements stay similar, but some become more pronounced and some become less. Sulfur, for example, stinks – but it's nothing compared with the stench of its homologues selenium or tellurium. Fluorine, on the other hand, is far more reactive than its homologues chlorine and bromine. Chemistry is, largely, about understanding how this works.

The problem is that as you increase the charge of the nucleus, some of the electrons – the part of an atom that controls its chemistry – gain velocity and begin to edge closer to the speed of light. That, thanks to Einstein's most famous rule, the theory of relativity, means they gain mass and change the radius of their orbit. It's a quirk called a 'relativistic effect'. It's why mercury is a liquid at room temperature, gold is golden, and lead–acid batteries work but tin–acid batteries don't. When it comes to the superheavy elements, the charge of the nucleus is so large that the trends the periodic table is supposed to follow – the whole point of deducing the properties of the elements by their position – may no longer apply.

The first superheavy elements, for the most part, behave as everyone expects. But once you get up to the outer reaches, things change. In the two decades since the first glimpse of 114, researchers have been probing, prodding and trying to perform chemistry on it. Because of its instability, typical lab experiments are out of the question (you can barely titrate a pipette before the atom decays or fissions). That meant Eichler had to do something fast and simple. When his experiment made element 114, instead of being caught by hitting something solid, it was imprisoned inside an inert gas filling a quartz bulb, flushed through a capillary (coated with Teflon so atoms wouldn't stick) and into an array of 16 selenium-coated detectors followed by 16 gold-plated ones. Eichler's kit was stuck out on the precarious balcony because it was the closest spot possible to the target – shorter beam line, less time wasted. Even so, from the moment an atom fused, to it running through Eichler's array, took about two seconds; by that time, it had probably decayed into copernicium (element 112).

Eichler's array had a temperature gradient, getting colder and colder the further along an atom travelled. Elements in copernicium's group all form amalgams with gold at different temperatures or occur in nature bound to selenium. By looking at which detector pings with an alpha decay, Eichler could tell at which temperature copernicium had formed a compound with either selenium or gold. Once he knew that, he could calculate the element's thermodynamics.

Eichler's experiment hints that something strange is going on. His results found he got pings at room temperature and at extreme colds. This, on its own, is unusual. But it becomes even weirder when you think about the basic rules of chemistry. Following a group of elements down the periodic table, you should get a neat trend for their thermodynamics when you plot the figures on a graph. Eichler's data show that both elements 114 and 112 don't appear to sit where they are supposed to on the curve for their respective groups. The research is only in its early stages – nothing about the superheavy elements is guaranteed until you repeat it – but it suggests that relativistic effects are already playing a part.

Jacklyn Gates at Berkeley sums Eichler's findings up neatly. 'We have a weird situation. Element 114 sticks to gold at room temperature, and it sticks to gold at near the temperature of liquid nitrogen, but not really in between. Those are two very different behaviours, and I don't think we have a good, coherent theory to explain it.'

This is exactly what Shaughnessy's team at Livermore is looking at. While element 114 is the extreme of what's possible to look at with modern chemistry, elements such as rutherfordium and seaborgium – once the far fringes of what was possible – are within reach, even if the experiments themselves are still tricky. 'Doing chemistry [on superheavy elements] is an extremely challenging enterprise,' Shaughnessy once told *Chemistry World*. 'There's theories that elements could actually alter their bonding based on what we'd predict just going down the periodic table. If that's the case, we could be revolutionising how we think of the periodic table.'

Shaughnessy's colleague (Padawan?) is John Despotopoulos. Young, with a scruffy stubble and long jet hair slicked back in a ponytail, he's one of the researchers looking at element 114 – and knows just how bizarre it can be. 'When you get to element 114, the chemistry is starting to behave more dissimilar to its direct homologue, lead, and behave more like mercury, which is in an entirely different group. By doing this chemistry, you might be changing the landscape of the periodic table.'

Despotopoulos is focused on the elements he can play with for longer than two seconds at a time. He's checking out lead and tin (both supposedly in element 114's group), as well as mercury (which isn't), looking at ways to separate them as fast and as accurately as possible. Currently he's creating traps for metals called 'thiacrown ethers', effectively cages made from carbon–sulfur rings. Again, the chemists are using their knowledge of the periodic table: sulfur is known for forming strong bonds with metals like lead and mercury. Once he perfects the technique, the next step is to try it on element 114. 'From that, you'd get some idea of whether it belongs in group 14 [with lead] or some of the more exotic predictions, like group 12 [with mercury]. Quantum mechanics … it's still just chemistry.'

Chemistry, yes – but chemistry that could potentially change how we think about the world.

* * *

On 30 May 2012 IUPAC agreed that there was enough evidence for elements 114 and 116 (the rest, it decided, needed a little more). Livermore and JINR agreed to split the names down the middle.

Element 114 was named 'flerovium'. To avoid any arguments about Flerov's relationship with Kurchatov and the Russian atomic bomb project, the team were careful to stress that it was named after the Flerov *Laboratory* as well as the man himself. Finally, Flerov had joined Seaborg on the periodic table. The two giants of twentieth-century element discovery – the men who had led their teams, led the world, into the unknown waters of nuclear instability – would be remembered forever.*

Thanks to a suggestion from Shaughnessy, so would Livermore. 'We chose the name "livermorium" for Lawrence Livermore National Laboratory, the city and all of the nuclear scientists that worked to make new elements,' Mark Stoyer recalls. 'We never could discover enough elements to name them after all of the important scientists.'

When Glenn Seaborg had phoned up the mayor of Berkeley to tell him that there would be an element named after his city, the man on the other end of the line hadn't been interested. When Mayor John Marchand found out the news Livermore would have its own element, it was the highlight of his career. Marchand is a chemist; he just got into politics to fix the local water supply.

Marchand and I are enjoying lunch, tucking into burnt ends on a table outside the stickiest rib joint in town, the effervescence from our soft drinks cooling the air. No need for 'I heart San Francisco' hoodies here; just golden sun to match the serene lustre of the hills. Directly across from us is a small park, wooden benches daubed with artwork from local graffiti artists. Each backboard celebrates a different aspect of chemistry – one depicts the Bohr model of the atom; another shows researchers with particle accelerators.

'That plaza is number 116,' Marchand tells me. 'So we named it Livermorium Plaza. I wanted public art there, so people could understand it's a special place. When the mayor of Dubna came out here to name Livermorium Plaza, he made a good point. There may come a time when Livermore or Dubna no longer exist; but as long as there is human knowledge, we will exist on the periodic table of elements.' The bond extends out into the district schools; in 2012, when the discovery team were awarded a $5,000 grant, they donated it to the Livermore High School

---

* While 'flerovium' had been suggested previously, nobody had actually used it, so the name was unlikely to cause any confusion.

chemistry department instead. Shaughnessy remembered the empty cabinets at El Segundo High; there was no way she was going to let the next kid potentially miss out on a career in science.

Dubna and Livermore are joined at the hip. The mayors have visited each other, proclamations have been made. And the elements have started to take on a very modern form of celebration. Marchand writes with a 'livermorium' pen; has a 'livermorium' tie; sports 'livermorium' pins. When the US team last went over to Dubna, they brought a present of wine glasses from the Livermore Valley. 'We rolled up to the Flerov Institute,' Marchand recalls, 'and we saw dials saying the temperature, and then gauges for background radiation. You go "OK, Toto, we're not in Kansas anymore." Then I'm sitting with Yuri Oganessian, and we're doing vodka shots out of Livermore wine glasses.' The Russians and Americans decided to christen the new elements with their own alcoholic beverages: JINR made own-brand vodka; Livermore made sure there was more than enough 'livermorium' wine to pass around. The local golf club even has its own tournament, the Livermorium Cup. After all, isn't it much more valuable per atom than gold, silver or bronze?

It's only with stories like this that you realise just how important the meeting between Hulet and Flerov was. Flerovium and livermorium aren't just important for chemistry; they have stitched communities on opposite sides of the world together.

Element 113 would do even more. It was about to unite an entire country.

CHAPTER NINETEEN

Beams of the Rising Sun

In November 1945 the US forces occupying Japan started dumping large metal objects into the waters of Tokyo Bay. Gathered on the deck of their ship, the sailors watched as these strange, alien creations were slowly edged off the side. One by one, the objects' lines were loosened and they began to topple overboard, hitting the water with a satisfying sploosh as they descended to their watery grave.

Back on dry land, Yoshio Nishina was distraught. The hunks of metal thrown into the sea were the remains of his cyclotrons. A month earlier, he had been granted permission to continue using them for medical, chemical and metallurgical research. But Robert Patterson, the US secretary of war, had changed his mind. Over a period of five days, working day and night, Nishina's machines had been ripped apart by engineers from the Eighth United States Army. A short time later, the engineers would also destroy cyclotrons in Osaka and Kyoto, even smashing a beta-spectrometer for good measure after misinterpreting a joke from one of the aghast scientists. It was pure vandalism: the destruction of every particle accelerator in Japan.

Nishina was one of the leading nuclear physicists in the world. In 1918 he had graduated from Tokyo Imperial University as an electrical engineer and had joined the Japanese Institute of Physical and Chemical Research (RIKEN). In 1921 he had been sent as a student to tour the research institutes of Europe, where he had become good friends with Niels Bohr. On his return to Japan, he had established his own laboratory and set out to probe the mysteries of the atom.

RIKEN was a unique set-up: a network that could be described as both independent research laboratories and a *zaibatsu* or business conglomerate. Founded in 1917 by scientists worried that Japan was losing step with the major powers, RIKEN's mission was, according to its architect Eiichi Shibusawa, to 'turn the country from imitation to creative power [...] to promote research in pure physics and chemistry'. Initially, the Japanese government had refused to back it, so instead it had been set up thanks to private donations, including some from the imperial family. For Nishina, RIKEN provided the perfect base for his research into quantum mechanics – and a backer with enough resources to help him discover a new element.

Japanese science had been trying to make its name in the world of element discovery since the start of the twentieth century. It had come close several times. In 1908 Masataka Ogawa, working under William Ramsay at University College London, had been testing a sample of thorianite when his chemical analysis came across something unknown. Ramsay, who had discovered the noble gases a few years earlier, encouraged the young chemist to publish his findings. Ogawa claimed he had discovered element 43, and called it 'nipponium' after his homeland. On his return to Japan, Ogawa had tried to follow up on his experiments – good science requires repetition, after all – but had been thwarted by a lack of modern equipment. Eventually, the claim was dismissed. Some modern researchers suspect Ogawa had discovered element 75 (today called rhenium) and misidentified it. If so, it was an easy mistake to make: both elements were in the same group in the periodic table and have similar chemical characteristics.

Nishina had also come close to discovering an element. Following Ernest Lawrence's blueprint, in 1937 he had built his own cyclotron, the first such machine made outside the US, which he had used to bombard thorium with fast neutrons, discovering the isotope uranium-237. This was a beta emitter and decayed into the then-undiscovered element 93. Nishina almost certainly created neptunium before Edwin McMillan and Phil Abelson confirmed their discovery – yet, like Ogawa before him, he hadn't been able to prove his new element.*

In April 1941, while Nishina was still trying to prove his discovery, he found his resources diverted. By then it had become clear that, for Japan to further its ambitions in the Pacific, war with the US was inevitable. RIKEN became occupied by project Ni-Go – one of the Japanese attempts to make a nuclear weapon. Nishina (the 'Ni' in the code name) was put in charge of the project and assigned himself the hardest task: enriching uranium.

Nishina had doubts that his homeland, with its lack of natural resources, could ever build an atomic bomb.† However, Ni-Go meant even more funding for his cyclotrons, so he played along. By 1944 RIKEN had created a 220t cyclotron that was a twin of its counterpart at Berkeley. Ostensibly it was to help make the weapon; in reality it was a research tool. Wisely, Nishina decided to keep the truth to himself.

Project Ni-Go collapsed soon after. In April 1945 RIKEN's main laboratories were bombed, destroying its thermal diffusion equipment. A month later, a German U-boat loaded with of 560kg (1,235lb) of uranium destined for Japan – a last, desperate throw of the dice by the Axis powers – was captured in the Atlantic. In June, Nishina told his superiors that the bomb project was over: a nuclear weapon was simply unfeasible.

The morning of 6 August 1945 changed his mind. A single bomb fell on the city of Hiroshima, levelling 12.2km^2

* It's a source of amusement to the Japanese team that neptunium – which could have been Nishina's – still ended up as 'Np' on the periodic table: it's the symbol that would have been used for 'nipponium'.

† As with the Allied effort, the generals in charge of Ni-Go didn't really grasp the concept of a nuclear bomb. On one occasion, Nishina's military liaison, Major General Nobuji, asked him why, if a bomb needed 10kg (22lb) of uranium, they couldn't just use 10kg of conventional explosives instead?

(4.7 miles^2) of the city and destroying almost 70 per cent of its buildings. Around 80,000 people died in the initial blast and immediate firestorm. A further 70,000 were injured, many with their clothes seared into their skin. Nishina was summoned to a secret government meeting where, despite the strict wartime censorship, he was shown a release from US President Truman. There, in front of worried officials, Nishina had confirmed Truman's claim that the one bomb 'had more power than 20,000 tons of TNT'.

It marked the end of the Empire of Japan. Heading to Hiroshima to survey the damage first-hand, Nishina left a note for one of his staff. 'If Truman is telling the truth, it is now the time for those involved in the Ni-Project to commit *hara-kiri* [ritual suicide].' Evidently, he reconsidered.

After the war, Nishina tried to preserve his cyclotrons, hoping they could help his country rebuild. As his creations fell into the sea, he knew he had nothing to offer. 'By the sad and untimely destruction,' he later wrote *in Bulletin of the Atomic Scientists*, '[the cyclotrons were] robbed of any chance to make contributions to science.' In Oak Ridge, the US scientists agreed, slamming the desecration of the machines as 'wanton and stupid to the point of constituting a crime against mankind'. Japanese science was dead in the water.

It was the least of the country's woes. For the first time in its history, Japan was an occupied country, its people starving, its culture and heritage reshaped and reforged by a victorious US. Rather than give up on his life's work, Nishina wrote to the Americans, asking for help to teach nuclear physics again. The response was blunt: 'All of Japan is hungry. If I were Japanese, I would take a shovel [and plant crops].'

Nishina ignored the advice and resumed his research. By the time of his death in 1951, Japan was on the road to recovering its lost scientific prestige. Yet it was still missing an element to call its own. RIKEN was on a mission to honour the legacies of Ogawa and Nishina. The emperor of Japan had helped establish the institute – and RIKEN wanted to repay the debt with a new element.

* * *

Tokyo is basking in a heatwave. The temperature is past 40 °C, but the hive never abates, never ceases and never stops. Commuters, packed like sardines, desperately fan themselves to stay cool; kids in school uniforms swoon even as they remain glued to their phone screens; waitresses dressed as French maids, dolphins or game characters try to lure passing customers into their cafes. Above and around, everywhere, blares electric activity. Anime creations wave at you from LED billboards, calling for your attention as a kaleidoscopic shower of stars cascades behind them. Trains course through the city's underground arteries in perfect synchronicity, rarely late or cancelled, while birdsong is piped onto the platforms to grant a moment's calm among the crowds. This is modern Japan – a blur of motion even the sweatbox heat can't slow.

At the far end of the Tokyo metro, the last stop on the sprawling tentacle of lines that make up the metropolis' subway, is Wakōshi Station. Here, the action gives way to a more sedate, suburban pace. Venture out of the station's south gate and look down. You'll see a bronze plaque emblazoned with an H: hydrogen. Further down the street is helium; then lithium; then beryllium. Keep following. The clattering pachinko arcades give way to sleepy suburban homes and neat company outposts. Eventually, you'll find yourself heading toward RIKEN's Wakō campus, home to its Nishina Center for Accelerator-Based Science.

Today, RIKEN is the largest research body in Japan, famed for pioneering work in areas such as pharmaceuticals, agriculture and neuroscience. Almost entirely government funded, its products appear in every corner shop of Japan, from energy drinks that make the body burn fat rather than carbohydrates to cosmetics based on amber. In 2010 the Nishina Center team used their accelerator to shoot carbon ions at cuttings from cherry blossom trees. The mutated blossoms – *Nishina otome* – bloom twice a year. In a country where cherry blossom viewing is a televised event, this is a big deal.

I'm not here to talk about any of those discoveries. I want to know why RIKEN joined the race to search for superheavy elements in the 1990s – and how it beat everyone else to the discovery of element 113. It's the element emblazoned on the last plaque on the walk from the station, a final marker that brings you to an abrupt stop outside RIKEN's gates.

'It looks like you've run out of space for plaques,' I say to my guide, Yukari Onishi, as she escorts me into the air-conditioned sanctuary of the main building. 'What happens if you discover another?'

'I don't know!' she laughs. 'I guess we'll have to lead them right up to the building. And after that start putting them indoors.'

Reminders of the hunt for elements are everywhere: in the foyer of the Nishina Center is a chart of the known nuclides created in Lego, the 3D model stretching up to show the instability of each isotope; with it, you can see the dip around the island of stability, teasing the element makers with its proximity.

In the US or Europe, discovering a new element barely registers on the evening news. In Japan, element discovery is followed as a national obsession. Hideto En'yo, the Nishina Center's director, remembers when the team detected one of the atoms that proved their claim. 'My daughter was in high school,' he recalls. 'I was about to visit and I told her I couldn't because something had happened. And she just said "Oh, you must have created an atom of element 113!" The high school students all know about our research.'

En'yo is youthful in appearance, his jet-black hair neatly combed, a broad smile on his face. He laughs long and often, more than happy to talk about the crowning achievement of his career. First, though, I need to present my gift. Business is ritualised in Japan, an elaborate and complex riddle of etiquette and social status where even bowing to the wrong degree can cause offence. When you present your business card (and you do, to everyone in the room in turn), you do so holding it on its corners, waiting for them to take it. When you receive a card, you read the name and keep it in a place of pride, never in your trousers. If someone is more important than you are, you always place your business card under their own. And when you visit a company for the first time, you try to bring a gift. As a visitor I'm not expected to do this, but it seems only polite to follow the local customs. En'yo takes my offering – a Royal Society of Chemistry cricket cap – with a smile. It seems the effort is appreciated.

'Discovering an element was a dream from Japan's history,' En'yo begins. 'A dream to recover from a mistake.' He's talking about Ogawa's 'nipponium' – failures sit heavy in Japan. 'And also Nishina, he tried to make a new element. He did it. OK, he couldn't confirm it, but if you judge him by present knowledge, clearly, he did it. He just didn't get the naming rights! For Japan, [discovering an element] has been a century-long project.'

The person the nation put its hopes on was Kōsuke Morita. There is even a manga comic about him, his cartoon alter ego imagined as a tubby figure with a bald pate and thick glasses. A nuclear physicist from Fukuoka, Morita left Kyushu University without completing his thesis (later insisting that he did not have the talent to finish it) and joined RIKEN as a researcher. In 1992 he went to Dubna, where he learned how to make elements under Yuri Oganessian. When it came time for Japan to enter the element discovery race, he was the natural choice to run the show. 'More than 30 years ago, Kōsuke Morita was charged to look into [element discovery],' En'yo explains. 'He needed 10 years to catch up with the world. Twenty years ago, we built the biggest atom smasher in the world. Then we were ready to compete. In 2003 we started the experiment – and we were going to win the game.'

En'yo isn't exaggerating. RIKEN's linear accelerator, RILAC, was easily capable of competing with GSI's own monster machine. It also has arguably the best detector in the world. But the team hadn't been able to procure calcium-48, and the machine wasn't set up to handle radioactive targets. While the Russians and Americans were racing ahead with hot fusion, the Japanese would have to use cold fusion instead.

It wasn't a bad call but this made discovery far harder. As the predicted reaction cross sections were much lower, fusion events would be much rarer than for the Dubna–Livermore group. But Morita's team had near-unlimited beam time. Element-making is like spinning a giant roulette wheel with a million numbers – spin it enough times and eventually your number will come up. If they ran their cold fusion experiment for long enough they were bound to get something. All they had to do was hope their results came before their rivals'.

It was a long shot.

In 2003 the Japanese team started bombarding bismuth targets with zinc-70 ions. In 2004 the team got their first hit: an isotope that decayed in 0.34 milliseconds. Even so, it seemed the race had been lost: six months earlier, the Dubna–Livermore group had already reported the discovery of element 113 from the alpha decay of element 115.

Yet neither team's claim was accepted immediately. The problem (for both teams) was that the alpha decay chains they reported broke down into undiscovered isotopes – meaning it was impossible to double-check if the results tallied with previous knowledge. There were also some inconsistencies with known data, probably due to the broad range of energies that elements with odd-numbered protons can produce. The IUPAC team arbitrating on element discovery ruled that both teams had produced 'very promising' evidence that was 'approximately contemporaneous'. However, it wasn't enough to say that the element had been made.

'The discovery was about who made element 113 without reasonable doubt,' En'yo explains. 'It's like an umpire: if they say "you win", the other side says "you're wrong". We were all left wondering … Dubna tried to directly create element 113, and they had two events. A month later, we had one event. They were faster than us, but they gave up.'

En'yo is half right. While the Dubna–Livermore group moved on, it was to focus on chemical experiments to shore up their discoveries – less taking a break, more gathering intelligence. 'In our minds, we had discovered two elements with one experiment – how economical! – and we were continuing to perform key experiments,' says Mark Stoyer. 'That is *not* giving up.'

Both teams continued to push, and by 2005 both had two direct 'hits'. It still wasn't enough to prove they had made the element. Going back to En'yo's umpire analogy, they needed another strike to end the game.

It didn't come for seven years.

* * *

In 1927 Thomas Parnell, at the University of Queensland in Brisbane, Australia, wanted to show his students that sometimes things that appear solid are, in fact, just really viscous liquids. Gathering his class, he heated a sample of pitch – the same stuff used to coat the bottom of ships – and plopped it in a sealed funnel. Three years later, he cut the neck of the funnel, placed the experiment outside the lecture hall and allowed the pitch to start flowing out of the bottom. The first drop fell five years later, in 1938.

Currently, the 'pitch drop' holds the world record for the longest continuously running lab experiment. It produces a single drop about once a decade – so far it's up to nine drops. The experiment (like Livermore's light bulb) is monitored constantly by webcam. John Mainstone, who inherited the experiment from Parnell, never saw the drop despite watching the pitch almost religiously for over 50 years. In 1988 he missed the rare event by minutes. 'I decided that I need a cup of tea or something like that, walked away, came back, and lo and behold it had dropped,' he told National Public Radio with a heavy heart. 'One becomes a bit philosophical about this.'

The pitch drop has nothing on RIKEN. Heating up pitch and leaving it alone isn't particularly expensive, and you can see when a drop is ready to fall for about a year in advance. RIKEN's hope involved blasting 6 trillion ions a second for months at a time at their rotating bismuth targets, hoping to see an unpredictable event that wouldn't even last a thousandth of a second.

RIKEN's control centre is organised chaos compared with the elegance of GSI or the industrial brutalism of Dubna; it feels like a cyberpunk lair. We've headed upstairs from the meeting room, into the workhorse section of the lab, away from the Lego models and into a realm of 24/7 science. Wires swarm out of circuit boards, monitors pile up and stained bins are filled with discarded energy drinks. The chairs are beaten and cosy. The whole place has a hum of sweat, perseverance and toil. Along the top of the control deck are two brightly coloured plush toys, a couple of creatures that look something like monkeys in space suits.

'They are Wakō City's mascots,' one of my hosts remarks. I forgot everything in Japan has its own mascot. Cities. Fire departments. Schools. Does RIKEN have a mascot?

Awkward silence. 'Uh … yes and no. There was one, once. It was a sort of, uh, termite.' A team member gives the Wakō mascots a reassuring pat. I get the distinct impression the space monkeys aren't going to be replaced any time soon.

I'm guided to a computer screen, a host of applications open on its desktop. To the side is a white block showing some kind of radioactive trace. In the centre of the screen is an immediate contrast: a black box with a red cross at its heart. It looks like an old computer game from the 1980s: no fancy graphics, just x marks the spot. 'This is element 113,' I'm told by my guide – they've called up the record of what they saw in 2012 to demonstrate what an event looks like. 'This white screen, here, means there's been an implantation event.' That's when a newly fused element ricochets off and implants on the detector. 'A red cross means you have an alpha-like event.' The ricochet has decayed. 'Three of those, you get a new element.'

It's hard to imagine how seeing that little cross – a single atom of element 113 – must have felt. It reminds me of the old video game *Desert Bus*, in which the player drives on a straight road for eight hours. Complete one trip and you get one point. It's so mind-numbing that gamers play it as an endurance feat to raise money for charity.* Most beam line scientists describe a single night running their machines as a tough ask: a crucible under which tempers fray and everybody slowly goes a little insane. At RIKEN, a team of 50 scientists spent a cumulative total of *553 days* of beam time just to see a little cross appear on a screen *3 times*.

They almost didn't succeed. By 2011 Morita's team had spent most of their budget, and the experiment was on the verge of being shut down. Bismuth and zinc are pretty cheap materials, but even so the team had burned through $3 million in electricity. There was also increased pressure to use RILAC for other, equally important experiments with a higher probability of success. Morita refused to back down. 'I was not prepared to give up,' he later said, 'as I believed that one day, if we persevered, luck would fall upon us again'. Morita would head to nearby shrines and temples in his spare time and pray, placing exactly 113 yen as an offering to the gods.

Then came an unlikely intervention. On 11 March 2011 the Tōhoku earthquake, the fourth largest quake ever recorded, shook Japan. It was the costliest disaster in history: almost 16,000 people were killed, almost 250,000 lost their homes and the destruction was valued at some $235 billion. In Fukushima, a nuclear power plant suffered three meltdowns, creating the largest nuclear incident since Chernobyl. In its aftermath, electricity prices across Japan skyrocketed. RIKEN's Nishina Center – which has eye-watering electricity bills at the best of times – went into effective shutdown. The element hunters saw an opportunity.

'It's a bit strange,' En'yo admits. 'The earthquake starved us of electricity, so there was great pressure not to do a lot. So, we said "OK, we just want to do one experiment." It meant we could run [the element search] most of the time. For two years, we shrank all of our operations except for the 113 search.'

---

* *Desert Bus* was initially created as a performance art piece by magicians Penn & Teller; even so, each year the marathon game session '*Desert Bus* for Hope' raises over $500,000.

In August 2012 the third event appeared. 'We had six alpha decays, seven after a beta decay,' En'yo recalls, thinking back to the red cross appearing on the monitor. 'And now there was no doubt about it – we had discovered element 113. We dedicated that third event to the people of Fukushima.'

The result, coming off the back of nine years of solid work, turned the Japanese team into overnight legends among the superheavy community. 'Imagine coming to work each 24-hour day for almost two years, and seeing no events on all but three days,' wrote Walter Loveland and David Morrissey in *Modern Nuclear Chemistry*. 'It requires an unusual degree of fortitude and courage.' Dawn Shaughnessy's praise is just as effusive: 'The Japanese team were pretty hardcore. I have nothing but mad respect for what they did.'

In 2015 the IUPAC working party met again. Both the Russian–American and Japanese teams had strengthened their cases. Researchers in Lund, Sweden, had confirmed the Dubna results, while the RIKEN group had directly synthesised new isotopes of bohrium, proving that it linked up with their recorded element 113 alpha decay chain.

It was a dead heat, and the element could have been awarded to either team. But ultimately, the working party found that the Dubna–Livermore claim hadn't fulfilled all criteria for discovering an element. The Russians were incensed: there was little doubt they had discovered the element first, and they had spent eight years and thousands of hours of beam time too. But the IUPAC decision was final – element 113 had been discovered by RIKEN.*

For 100 years, Japan had dreamed of an element on the periodic table. Finally, it had one. The celebration was so large it was even attended by Crown Prince Naruhito, honouring the

---

* To make matters worse, the IUPAC finding was littered with technical errors; to this day, the Dubna–Livermore group feel robbed. 'Details matter, and the sloppy IUPAC report is unsatisfactory,' Mark Stoyer notes. 'I think all the hard-working scientists in this field have again been done a disservice.'

imperial family's long-standing connections to RIKEN. 'I am deeply moved by the addition of the new element,' he stated – before observing that he used to copy out the periodic table by hand in high school. In Japan, there could be no higher praise.

Dubna and Livermore had toasted their successful discovery of 114 and 116 with 'flerovium' vodka and 'livermorium' wine. RIKEN went a little further. In 2010 the Nishina Center team had put a batch of brewing yeast into their RILAC beam and induced mutagenesis – changing its genetic code to create an entirely new strain. The result was *Nishina Homare* sake ('in honour of Nishina'). What better way to celebrate an element than mutant rice wine created from your own ion cannon?

Perhaps the most cathartic moment was the choice of name. 'Nipponium' was out of the question – Ogawa had already used it for his misidentified 'element 43', and IUPAC's rules were clear that a name couldn't be repeated. But there are two words in Japanese for their homeland, the land of the Rising Sun: *Nippon* and *Nihon*. Element 113 became 'nihonium'.

★ ★ ★

'Why produce new elements and isotopes? It's a good question. They have short half-lives and no practical application.' Hiromitsu Haba, one of the RIKEN chemists, takes his time as he thinks about the answer. Haba is one of the team who dedicated over a decade of their lives to hunting down element 113. What makes that level of commitment worth it?

'The elements are very important for the universe, for the body, for everything!' Haba says finally. 'If we can understand such elemental particles, we can come up with good theories. Currently, we know 3,000 isotopes. But, theoretically, there are 10,000 isotopes. We only know a third of our world.'

Haba is referring to the latest models. Since Maria Goeppert Mayer and Hans Jensen blew the understanding of the nucleus wide open with their shell model, physicists have been trying to work out just how far the periodic table stretches – how much of the building blocks of existence remain undiscovered. Usually, this is presented as the chart of nuclides – like RIKEN's Lego model or Glenn Seaborg and Georgy Flerov's drawings of the 'sea of instability'. The borders of this map are the 'drip lines': beyond the neutron drip line, the nucleus kicks out a neutron before it forms; beyond the proton drip line, the same happens for protons. Anything between the drip lines is theoretically possible. Currently, the best guess – and it is only a guess – is that the elements as we know them stretch out to number 172.

RIKEN is hard at work to fill in these blanks. Between 2016 and 2018, the team discovered 73 new nuclides, from isotopes of manganese to erbium. Each contained more neutrons than ever before seen. All of them were created using fission. Rather than trying to avoid the element splitting apart, the Japanese team have revelled in it, blasting uranium at a target made of beryllium – one of the lightest elements – in the hope that the uranium atoms would break into interesting fragments.

While these isotopes may seem pointless, Haba is quick to point out that history says otherwise. 'Technetium was the first human element made,' he says, thinking back to how Emilio Segrè found element 43 from one of Ernest Lawrence's leftovers. At first, it didn't seem too interesting. 'Now, technetium is very important for nuclear medicine. Every year, 1 million people use radioisotopes in Japan … [and] the lanthanides are used in magnets or mobile phones. Nobody knew they would be used that way at the time they were discovered. Each element is similar, but each has its own use. Neodymium and lanthanum are similar, but they have their own uses … element 113 [and the other superheavy elements] may have a use too.'

One of Haba's interests is the superheavy element seaborgium. As with Robert Eichler, Haba and his colleagues are doing rapid-fire chemistry experiments to see how their fleeting products work. RIKEN even have robots to control the process, buying them valuable half-seconds in the world's fastest chemistry experiments. 'This is an example,' Haba says, bringing up a molecular structure on his computer. The seaborgium atom is in the centre of a six-pointed, three-dimensional star. At each point is carbon, then oxygen. It's a classic chemical structure known as a hexacarbonyl. 'We produced two seaborgium isotopes, separated them and caught them. Then we added carbon monoxide (CO) here, so we can see if it interacts. Now we know that this hexacarbonyl compound exists. By heating the molecules, we can destroy them and investigate the bond strength between the seaborgium and carbon. We can then compare them with theoretical calculations.'

All of this is again part of rewriting the periodic table. Seaborgium is, supposedly, in the same group as tungsten. But what if it doesn't behave like tungsten at all? 'The structure of the periodic table is not going to change,' Haba stresses. 'The element is put on the periodic table irrelevant of its properties ... but it's very difficult to get used to the chemistry on this row of the periodic table.'

Not everyone agrees. As with the weight of the kilogram, science has a habit of self-correcting. While an element's number on the periodic table is static, positions can move about: after all, until Glenn Seaborg came along, uranium had been placed under tungsten, the very position seaborgium occupies today. 'As a chemist, the usefulness of the periodic table is the periodicity – if shown that these new elements belong in a different group, they should be moved there,' observes Nancy Stoyer. 'The periodic table is a *living* construct.'

If this debate sounds pointless, it's anything but. By working out how the relativistic effects change the elements, how they stop following rules that science has trusted for centuries, we can work smarter and find new ways to use the elements we have discovered. Remember the search for naturally occurring superheavy elements in the 1970s, with the US and Russian teams launching expeditions into hot springs or the depths of the Gobi Desert to try and find them? Today's researchers know those searches were looking in all the wrong places: they were basing their hunt on incorrect assumptions about how superheavy elements behave. Years were wasted because we didn't understand the rules of physical reality.

It's only by discovering more elements that we can work out what those rules really are.

★ ★ ★

RIKEN – like the rest of Japan – isn't satisfied with one element. Already, its researchers are hunting for more. RILAC is being reconfigured for new experiments; the Nishina Center's oldest cyclotron has already started the search. Both machines are going to run in parallel to hunt for elements 119 and 120 until they are found. With the new elements' cross sections predicted to be orders of magnitude lower than nihonium, there's no point trying cold fusion. Three hits could take centuries. Instead, the RIKEN team have taken a leaf from Oganessian's playbook and switched to hot fusion.

The reconfiguration of the linear accelerator is the big challenge. It has already cost $40 million just to make the changes required. 'Not all of the linear accelerator's parts are superconducting,' En'yo explains. 'That lets you go to a lower charge state more effectively. To convert it requires two years, which is why we decided to use the cyclotron as well. The cyclotron isn't better than the linear accelerator, but with a good ion source we can overcome that and get a reasonable enough intensity to start the 119 search until the linear accelerator is remodelled with a stronger beam.'

This revamped machine will bring new demands. When the team discovered nihonium, shooting zinc into bismuth, the main cost was electricity: zinc and bismuth are cheap. Conversely, hot fusion requires curium targets, created and shipped over especially from Oak Ridge's HFIR. The search will cost $1 million a year. But it doesn't matter if the new experiment costs millions or takes another nine years to produce its success: as it has proved time and time again, RIKEN doesn't back down.

'We'll keep running the experiment until we make the discovery,' En'yo says. 'Or someone else does.'

The someone else is Yuri Oganessian. While the Japanese were searching for nihonium, he had finished the seventh row of the periodic table.

CHAPTER TWENTY

The Edge of the Unknown

Commercial airlines carry all kinds of curios in their cargo holds. The average passenger flight has a couple of tonnes of freight on board: everything from the post and pets (several million animals fly around the world each year), to rarer items such as live lobsters for restaurants or stacks of gold bullion. In 2012 a crocodile being transported from Brisbane to Melbourne escaped its container and roamed the cargo hold of a Qantas jet until it was discovered by the baggage handlers. Generally, if something needs to arrive in a hurry – be it a heart on ice for a transplant or a sample of horse semen for breeding – a commercial airline is the way to go.

The person with the final say over what comes on board a plane is the captain. With almost complete power, a pilot can decide whether to refuse a passenger, turn away a piece of cargo going in the hold or even abort the flight entirely.

In 2009 an unusual package made its way to JFK Airport, New York, and onto a Delta flight bound for Moscow. As the captain checked the manifest, he raised an eyebrow at the strange cargo: a piece of metal about the weight of a sesame seed that was encased in lead and plastered with radioactive warnings. It was a sample of element 97, berkelium. And this was the fifth time it was about to cross the Atlantic Ocean.

For years, Yuri Oganessian had been trying to make element 117. All the other 'missing' elements of the seventh row – 118, 115 and 113 – had, by then, solid evidence for their existence. The problem was that if the team were to stick with a calcium-48 beam, they needed a berkelium target. There just wasn't enough of it to make one.

As mentioned in Chapter 3, there are only two places on Earth capable of producing large quantities of berkelium: the Research Institute of Atomic Reactors at Dimitrovgrad, and HFIR at Oak Ridge. As berkelium has no commercial use, neither the US nor the Russians had any reason to make it directly. Instead, it was usually found as a by-product of its daughter element, californium, which could be used to start up nuclear reactors, identify gold and determine the geology of oil wells. In stark contrast to its neighbour on the periodic table, in the 60 years since it was first created californium had become the most valuable metal on the planet. Today, the asking price is around $27 million a gram.

Oganessian repeatedly tried to convince Dimitrovgrad to supply any leftover berkelium from a californium production run, but had no success. Instead, he turned his attention to Oak Ridge.

In 2005 Oganessian contacted Joe Hamilton at Vanderbilt University, based in Nashville. The duo had worked together for nearly 20 years and had published 200 papers together. Hamilton, not giving a damn about the Cold War, had made his first trip to the USSR in 1959 and had spent more than six months of his life in Russia. He also had ties to Oak Ridge. 'We went and talked to the people at the High Flux Isotope Reactor,' recalls Hamilton. 'They said the only economical way to get berkelium was to piggyback a commercial company order for californium-252.' There was only one problem: there was no commercial order for californium-252 at the time.

Hamilton didn't give up. He phoned Oak Ridge every three months for three years. His tenacity became the stuff of legend. 'He pounded on the door,' remembers Mark Stoyer. '"Can we get berkelium? Can we get berkelium?"'

In August 2008 Hamilton's dedication finally paid off: there was a californium campaign scheduled at the reactor. The next month, Oganessian flew over to Vanderbilt to celebrate Hamilton's 50 years of research in atomic physics. There, among the canapés and polite conversation, the American told him the good news. Still finishing their lunch, they invited Oak Ridge's associate director, James Roberto, to join them. Roberto listened to their plan for the element and realised its potential. Soon, Oak Ridge, JINR, Vanderbilt University and Livermore (contributing some of the costs of making berkelium) and the University of Tennessee, Knoxville were working together.

In December 2008 a californium run at HFIR in Oak Ridge produced the 22mg of berkelium. The teams were on the clock: the sample had a half-life of 327 days. It took 90 days to cool the sample, then 90 days to purify it. Next, it had to be prepared for transport. First, it was dried into a solid form (berkelium nitrate) and slipped into a glass vial. This was swaddled in chemical wipes (for shock absorption) and placed inside a lead pig – a carrying container about the size of a coffee flask. Finally, to complete the package, the pig was sealed and placed inside a drum, which was sealed again. 'And that's your Department of Transportation certification,' Oak Ridge's Julie Ezold told me. 'Then it's literally like FedEx.'

At least, it should have been like FedEx. 'In this particular case, since it was going to Moscow, it went to JFK Airport, New York. It flies over to Moscow. And [the group] had spent *months* trying to get the documentation prepared for this. The documentation didn't make it on the airplane. So, it came back. We fixed that, and it goes back over again … and something in customs wasn't right. Again, it comes back. So, we are very concerned, fighting the half-life, of how many trips this took across the ocean.'

The fifth flight proved lucky – after 25,000 air miles, the berkelium made it to Dimitrovgrad, where the target discs were to be prepared before heading on to JINR. Even then, a Russian customs officer tried to open the package and check the contents. Luckily, some of the Dubna team were on hand to convince the official he *really didn't* want to see what was inside.

The Delta pilots who shipped the vial, and the Russian customs officers who tried to check them, probably don't know they helped to write the final chapter to date of chemical element discovery. After firing their calcium-48 beam into the berkelium for 150 days, the Dubna–Livermore team announced 6 atoms of element 117 in April 2010. In 2012 they did it again. In 2014 GSI confirmed their data. Element 117 had been found. Its half-life was a mere 50 milliseconds, but this was still far longer than would have been predicted if the island of stability were a myth. The new element's alpha decay chain matched up with the results from the search for 115 too, confirming the validity of the experiment.

In 2012, when the experiment was repeated, the team got a surprise. A small amount of their berkelium target had beta-decayed into californium. As well as producing yet more atoms of 117, the team also hit one of these rogue spots of californium – resulting in yet another atom of element 118. It had a half-life of less than a millisecond – but it was yet more evidence that Livermore and Dubna, with a little help from Oak Ridge, were the new masters of creating the different components that make our physical universe.

In December 2015, when the IUPAC/IUPAP joint working party announced that element 113 had been granted to RIKEN, they also announced that elements 115, 117 and 118 had been discovered by the Dubna-led team, with 118 being awarded to Dubna and Livermore.*

Yuri Oganessian had led a collaboration that stitched together 72 scientists from 6 institutions around the world. The team had been awarded the discovery of five elements. The seventh row of the periodic table, something that had remained partially complete for more than a century, had finally been filled.

On 23 March 2016 the collaborators held a conference call to discuss the names of the new elements. First, 117: it would be 'tennessine', honouring Oak Ridge and the contributions that had come from across the entire state. Next, 115: 'moscovium', after Dubna's home state.

Livermore had wine. Dubna had vodka. RIKEN had sake. When Oak Ridge found it had a new element, lead

* It would be convenient if the elements had been discovered in numerical order, but they weren't. In summary, the order was: 114, 116, 115, 113, 118 and 117. Don't be surprised if element 120 is discovered before element 119.

physicist Krzysztof Rykaczewski popped to the Jack Daniel's distillery in Lynchburg and ordered up some Tennessee sippin' whiskey. Ask Ezold, Rykaczewski or any of the Oak Ridge scientists the proudest achievement in their career, and they'll tell you it's discovering a new element. It's not just for them, either. 'I have an eight-year-old daughter,' Ezold told the *Farragut Press* when the names were announced. 'For me to be able to say I was part of the discovery of a new element is off the charts.' Powering space probes or helping to land a rover on Mars doesn't even come close to being a mum who helped make and name a piece of reality.

Do names really matter? Again, Sigurd Hofmann's answer is perhaps the best. 'Partly, I suppose it is because everyone feels qualified to give an opinion, and partly because they become very proprietary and emotionally involved. It is, after all, the very centre of the lives of people doing it.'

Element 118 was about to become the very centre of one man's life in a way he could never have imagined. After tennessine had been named, Yuri Oganessian was asked to step out of the room. Then, with every other scientist's agreement, the Dubna–Livermore collaboration decided he would follow in the footsteps of Glenn Seaborg. Element 118 would be 'oganesson'.

'Phew.' Oganessian blows out his cheeks when he thinks about the honour. His eyes widen with emotion, pupils blinking with a thin film of tears. 'It's difficult to say how I feel. Because it was proposed by my collaborators. All of these people were involved and came to this conclusion after all the elements we synthesised. Of course, it's an honour for me. But it's my friends and colleagues who want to express it. I'm just seeing that there are a lot of things yet to do, even with this element. It's not the final piece of the story.'

Some people dream a little bigger. And, like the Japanese, Oganessian hasn't finished dreaming yet. He's going after elements 119 and 120 too. And this time he has even more help.

* * *

It's easy to get the impression that element discovery is only about the labs where the elements are made. This is a complete illusion. The elements were made in Berkeley, Dubna, GSI and RIKEN – but without partnerships or the flow of personnel and ideas, we'd still be stuck with a periodic table ending in uranium.

Today, the superheavy world isn't just defined by discovery. There are a multitude of other labs, in France, Japan, the US, the UK, Sweden and Poland, where outstanding work is being done by brilliant scientists to contribute to a greater whole. It's part of the reason I've made the longest leg of my superheavy journey, all the way to see an accelerator in Canberra, Australia. Here, one team is trying to map out the path to the undiscovered elements – 119, 120 and beyond. I just hadn't counted on the Aussie weather: Australian National University (ANU) is flooded.

Canberra is an artificial city, a compromise capital to stop Sydney and Melbourne arguing about who is better. Set in lush hills and split by an artificial lake, Canberra's manicured lawns and far-too-organised streets feel over-engineered compared with the organic blossoms of Berkeley or Dubna. Yet unfortunately for the Australian government, even a city planned to perfection is at the mercy of the weather. Last night, as I fought off jet lag from a 24-hour flight, 2 months of rain fell in a matter of hours, bursting Sullivans Creek and dousing the ANU campus in rainwater. Officially, everything is closed – even the redback spiders have scurried for higher ground. But beam time is too precious to let a little thing like a deluge stop the work. The ANU accelerator is up and running – and that means I can make my visit.

Despite the university supposedly down to a skeleton staff, almost every single member of the department has come in to meet me. The ANU team are close-knit, a mix of young and old sharing a small office building on the fringes of campus, just a short walk away from their particle accelerator. Camaraderie is high: when three of the professors wanted to give their technical staff more opportunities, they pooled their own money to create a fund to get them to overseas conferences. The relaxed atmosphere extends to the break room. Here, visitors can find two fridges adorned with signatures of departing visitors. 'Those are our beer fridges,' one team member tells me, as if it's the most natural thing in the world. 'When you leave, you've got to try and fill the fridge with beer. If you manage to do it, you get to sign your name.' Welcome to science the Australian way.*

The relaxed, friendly feeling continues when you go across to the accelerator itself. The control centre is a single room with beaten sofas and an old-school accelerator control console first used in 1968. Even so, you soon realise just how staggeringly complex it is to run, fund and house even a small heavy element facility. The ion accelerator – that first push for your particle cannon – is kept vertically in a tower, dropping down a 33m (108ft) tank filled with 20t of sulfur hexafluoride gas to prevent the 14MV of charge sparking off against the wall. Shooting down at around 35,000km (22,000 miles) per hour, the beam is then bent by magnets into the horizontal plane (using the Lorentz force, the same way a cyclotron's beam is bent) and either directed out for experiments or – if they need the extra kick – into the superconducting linear accelerator itself, where it's sped up using the same beam pulses found at GSI. The shape of the accelerator is a little different, though: cramped for space, the ANU team have bent it around in several loops, the beam line twisting its way to take up virtually every spare inch of the accelerator hall before swinging out toward its target. It's like a game of *Snake*, in which you're forced to either climb over the pipes or scramble under them to get to the other side of the lab.

'Some people said it was cursed,' remarks David Hinde as we step up above the linear accelerator. The director of the

* Despite the name, the fridges usually contain packed lunches; the team don't drink while operating the particle accelerator these days – although there used to be a monitor next to the fridge to check on the beam …

Heavy Ion Accelerator Facility, he's leading my tour with experimental physicist Nanda Dasgupta. 'The accelerator came originally from Daresbury [a nuclear physics laboratory in the UK], and before that Oxford University, after they took a funding hit. We were fortunate to get it, but everyone who had it 'fell over' [ran out of funding] – it had been at two different labs but had never really been used. Anyway … the beam comes down from the tower, bends into a horizontal plane, around a bend – which is *preeeeety* difficult, very particular entry, very acrobatic – and then it comes in here …'

At a larger laboratory, the chemists and physicists are often hands-off with the equipment – they just worry what's going on with their beam line. Without the funding for technicians to run the accelerator around the clock, Hinde, Dasgupta and their students adjust the beams, magnets and ion source themselves. The upside, Dasgupta says, is that you can fiddle around and make small adjustments to perfect your set-up on the fly. 'If you see something unusual, you can just change it,' she enthuses. 'Last week I was in the tea room and someone said, "Oh, it'd be good to measure this, when can you do it?" And I told them: "Tomorrow!"'

The ANU mantra is to be low-cost but high-precision. Dasgupta passes me a metal square frame. Inside, a $1cm^2$ circle is filled with graph paper. Holding it up to the room's fluorescent lights, I can see a pinprick hole burned through it. 'That's the size of the beam spot,' she says. A little further on, we come to the experimental equipment at the end of the beam lines, including a Faraday cup to catch the beam. It's a mere 4.5mm in diameter. The fusion detectors, designed at ANU, use thousands of 20-micrometre tungsten wires, coated in super-high-quality solid gold. They've been shipped from Norway, where they had originally been made for CERN. Once again, the Canberra team had to make the equipment themselves, although Hinde didn't mind. 'I was of the Airfix generation,' he says, wiggling his eyebrows. 'I used to make Second World War airplanes and tanks. When someone tells me you can't do something a certain size, I say "sure you can".' If you can stick a rear-view mirror into the cockpit of a scale model Spitfire, you can make a fusion detector.

The set-up isn't as powerful as the leading labs' machines. Its kit can't reach the staggering sensitivity needed to get down to picobarns. But ANU doesn't need it. It might be small and on a shoestring compared with the bigger labs, but it still punches above its weight. Currently, it's a world leader on quasi-fission – providing essential clues about why nuclei generally fail to fuse together. 'We're known for our fusion work,' says Dasgupta. 'We also look at tunnelling, well below the fusion barrier.' This is essential for fusion in stars, where the energy provided by pressure and temperature isn't enough to push past the Coulomb barrier. The trick is known as quantum tunnelling: when you hit the subatomic levels, things get weird and start acting like waves, allowing particles to slip through.*

While ANU's experiments can't create new superheavy elements – there's not enough beam intensity and the university lacks the multi-million-dollar infrastructure needed to handle highly radioactive targets – it does mean the scientists can provide clues about where and how to look for them. The Australian team use their accelerated beams to work out the best ways to form superheavy elements: what energies are needed and how the reactions will behave. I remember Jacklyn Gates talking about magnets being just 6 per cent out of alignment; ANU's work means the element hunters now have a better idea about what beams and targets to try, and where to look. 'It's like *Where's Wally*,' Dasgupta explains. 'Wally is hidden somewhere in a load of pirates. We're getting rid of the pirates, taking Wally and putting him in the detector.'

Thanks to the work by Hinde and Dasgupta, we understand how stable and unstable nuclei work, why our world can cling

---

* Quantum tunnelling is a key part of understanding the atom, but delves into the strange world of quantum physics and is too complex to go into here.

together. It also provides clues as to how our world changes when we move from the subatomic, quantum world with its own rules and forces, through to the classical physics we're more comfortable and familiar with.

More importantly, they help with games of 'element *Where's Wally*'. And when it comes to the hunt for new elements, scientists need all the help they can get.

★ ★ ★

In 2013 sitcom character Sheldon Cooper figured out a way to synthesise element 120. Less than 10 minutes later (not including ad breaks), a team in China used a cyclotron to confirm his discovery. Then, Amy Farrah Fowler, Cooper's equally fictional girlfriend, pointed out that he'd made an error by a factor of 10,000. 'I'm not a genius, I'm a fraud,' Cooper whined. 'Worse than a fraud, I'm practically a biologist.' Despite Cooper's protests, the world of *The Big Bang Theory* is still doing one better than reality when it comes to discovering elements.

The funny thing, at least for nuclear scientists, is the sprawl of calculations on Cooper's whiteboard. Every single one of them is a plausible way to create element 120. And every single one of them (except Cooper's supposed breakthrough using mendelevium – its half-life of 51 days means it just doesn't hang around long enough to make a good target) has been tried in real life. So far, no new element.

The big problem with finding element 120, or even 119, is that there's no immediately obvious way to do it. While calcium-48 is an amazing, neutron-rich beam, to make element 119 you'd need a target made from einsteinium, which is currently only produced in nanograms at a time. 'In principle, it's not impossible,' insists Oak Ridge's James Roberto. 'If we were able to take a reactor and run it in a very special mode and seed it with a gram of californium – and all of those are multi-million-dollar ifs – we could make about 40μg. That would be 0.1 centimetre square of a target, about the size of the beam itself. The challenge is that if we could get this project together, you'd still need to make a target that could take the whole beam in one spot for months at a time [without burning up].'

It's a lot of money for a lot of uncertainty. RIKEN is currently shooting vanadium into curium. The other promising idea, which the Dubna–Livermore group is working on, is to use a titanium-50 beam (six extra neutrons), shot into berkelium. Neither option is as good as a calcium-48 beam with its eight extra neutrons; the only question is which team's approach will succeed first.

Both teams are getting the material for their targets from Oak Ridge. But nobody knows which team is likely to push the superheavy elements onto the eighth row of the periodic table first. 'We're not quite stuck yet,' Roberto says, 'but the breakthrough is not obvious.'

The options to make the next elements don't end with Oak Ridge's reactor, RIKEN or at the bottom of Sheldon Cooper's whiteboard. In 2011 Sigurd Hofmann's GSI team tried to find element 120 using a chromium beam into curium. At around 4.20 a.m. on 18 May, they saw ... they don't know. 'We saw *something*,' Hofmann hedges. 'Part of a decay chain that agrees with data from Dubna in around 2000. But later these data were no longer mentioned. So far, it's not confirmed.' As always, the world of element hunting has lots of possible routes to success; it's backing them up that's the hard part.

There are other ideas. Roberto has suggested reversing the reaction, building the target out of the lighter element (for example, iron) and shooting it with a beam using plutonium ions. It's possible, although the beam intensity will be far lower, making finding an element that much harder. It also has a big drawback: shooting plutonium through an accelerator would contaminate it with radiation for 1,000 years. 'It's not science fiction,' Rykaczewski promises. 'You just need a reasonable accelerator that's closing down ... then it's the perfect experiment.' Another option is to shoot two heavy nuclei into each other, such as throwing curium into uranium and hoping that as the two bounce off a few protons and neutrons get transferred in passing. It's a technique called multi-nucleon transfer. Again, fine in theory – and could even provide a bridge to the island of stability if you used isotopes with enough neutrons. Sadly, it has a major drawback: the amount of fission noise you'd get from the reaction would make it virtually impossible to see (and more importantly, prove) you've made your new element. It's another reason why work from the likes of ANU is so critical.

For now, the element hunters seem to have a path forward. Everybody expects that elements 119 and 120 will emerge in the next five years. Beyond that, nobody knows. And that raises one last question. How far does the periodic table go before it breaks … or are we there already?

CHAPTER TWENTY ONE

Beyond Superheavy

In 2017 a paper emerged from Massey University in New Zealand and Michigan State University in the US. It marked the end of chemistry as we know it.

Element 118, oganesson, it claimed, does not have electron shells.

Oganesson sits in the group of the periodic table occupied by the noble gases: the likes of helium, neon and argon. These, chemistry students are taught, have filled electron shells, meaning they don't usually interact with things. But using state-of-the-art simulations, the New Zealand and US team compared the classic, expected appearance of the noble gases, and then a model that took into account relativistic effects. As the nuclei became heavier, the effects became more pronounced. By the time the model reached oganesson, the supposed electron shells are more like electron soup. It's called a Fermi gas, after the man who first imagined it could exist. Even today, Enrico Fermi's brilliant mind is being proved right.

If true, oganesson means the periodic table stops being relevant in terms of predicting properties. Its neat rows and patterns break. The electrons would be easy to polarise, meaning that oganesson would be reactive. It would form compounds with other elements and molecules easily. It wouldn't even be a gas at room temperature – it would probably be a solid. Sadly, only five or so atoms of oganesson have ever been created, and it is so unstable it lasts for less than a millisecond. That's too small an amount, and too little time, to test the theory.

One of the authors of the oganesson paper was Witold Nazarewicz. He's one of the world's leading physicists, who has gone from a PhD in his native Poland to chief scientist of Michigan State University's Facility for Rare Isotope Beams.

When it starts up in 2021, the hope is that it will be able to produce neutron-rich radioactive beams that will allow researchers to find a way to get closer to the island of stability.

Nazarewicz says the strangeness around oganesson is only the beginning. Eventually, the periodic table is probably going to collapse. 'We can only calculate nuclear properties using the best current models,' Nazarewicz explains. These are the same models that first said the periodic table would end at element 100; that the nucleus was like a drop of water; that it had shells and magic numbers. The models change according to the best evidence available, with each iteration becoming more accurate and precise – even if it takes us far outside our comfort zone. 'We can calculate,' Nazarewicz adds. 'And theorists should be allowed to calculate things that are very exotic.'

It's these calculations that turn everything on its head. Using a combination of nuclear surface effects, quantum mechanics and Coulomb repulsion, Nazarewicz and his colleagues are mapping out the shape a nucleus will twist itself into while trying to hold onto its protons. There are a host of ideas about how the heaviest atoms may contort – nuclei stretching out, folding in on themselves, even warping into a doughnut shape with a hole in the middle. By the time you get to the elements around 140 or so, the latest models suggest that things become very strange. 'The heavier you are, the more unstable [things become],' Nazarewicz says. 'Coulomb repulsion means you might have to deal with exotic topologies. Proton density might create a hole, or even a void. We don't know. There will be nuclides – proton–neutron cluster systems – but they might not have electrons.'

It's an astonishing claim. The very definition of a new element is based on the time a nucleus takes to attract electrons. Instead of having new elements, you'd suddenly have a section of the periodic table that would be … blank?

'There would be nothing!' Nazarewicz confirms. 'There would be missing windows, there would be gaps [in the table]. Elements are basically for chemists. There's no chemistry without electrons. You'll have a nucleus but *no atom*. The end of chemistry. It might be that chemistry will die, and then suddenly, something may appear with extra stability. And if it lives a long time, and it attracts electrons ... Hey! Chemistry is reborn! It's so fantastic that one is afraid to think about it. But it might be there!'

Don't worry just yet: the periodic table isn't going anywhere. These models are predictions far ahead of where we are in terms of element hunting and might not even apply – at the moment, they are just theory. Eric Scerri, a chemist and philosopher of science at the University of California, Los Angeles, thinks the periodic table is safe for now. 'It's still going to be relevant,' he argues. 'I'm reluctant to make changes. Relativistic effects change things, but when it comes to how to organise the periodic table, I don't think we should build in any major contortions. It's an approximation.'

The periodic table, as Scerri points out, bases itself on more than just an element's properties. Modern chemists follow something called the Madelung rule, which uses energy levels to predict which subshells orbiting an atom an electron will join. The rule has its exceptions, but is still the most general way we have to deduce how the periodic table as a whole should be organised.

For elements 119 and 120, it's obvious where they should go: at the bottom of group 1 and group 2, kicking off a new row of the periodic table. The problem is element 121 and beyond. Here, in the rows above, are where the lanthanides and actinides start – those elements sat in their own section at the bottom of the periodic table. If you follow the standard pattern of how the elements are laid out, 121 would sit under the actinides – a 'super actinide' group. The table would then follow the pattern it's always done. According to the latest definitions, anything heavier than element 126 will be a 'beyond superheavy' element.

It's not the only option, however. Pekka Pyykkö, from the University of Helsinki, is one of the leading theorists on where the periodic table will end. His own model has 121, starting another box cast off from the main periodic table, split away from even the lanthanides and actinides. This would involve an entirely new electron shell type – very compact and deep inside the atom. For Pyykkö, elements 139 and 140 would be part of the main table, while element 141 onwards would form a new row placed under the actinides. 'It's going to be a mess,' Pyykkö admits. 'But this will give you, from a chemical point of view, a possible periodic table.'

The problem with any of these models is that nobody really knows what's going to happen. We don't know where the periodic table will end. We don't know how it should be arranged. We don't even know if our current table is right. Currently, the best models say that there are 172 elements. Some argue for 173; others fewer. Some physicists don't see why we should even stop at all. Ask three theorists, and you'll get three different theories.

'Element 172 is a chemist's calculation,' points out Mark Stoyer. 'But once you reach the massive size of an atom that element 172 would be, the innermost electrons are very relativistic. They gain mass, so are heavier, so the orbit contracts so much those electrons spend a significant fraction of time *actually inside the nucleus*. Amazing when you think about it – electrons whizzing around inside the nucleus.'

Stoyer's right – it would be loosely the equivalent of the Earth orbiting through the Sun. This is a problem even the world's fastest computers, such as Oak Ridge's Summit, can't solve. 'We can completely calculate effects only for very light nuclei,' Stoyer continues. 'After carbon [element 6] we start to run out of steam. Anything larger is intractable, even given the biggest supercomputer. This makes superheavy elements so fascinating!'

For now, the debates will remain theoretical. Already, we've come a long way from Antoine Lavoisier's list of elements, or the table as put together by Dmitri Mendeleev or Henry Moseley. As Nancy Stoyer pointed out earlier, the periodic table is a living construct. It is constantly made, broken and made again. While we think of it as relatively static because it sits on our wall, in truth it's a malleable tool: it's a guidebook to our chemical universe – and guidebooks get updated.

'And what if Nazarewicz is right?' I ask Scerri. What if there is a region of the periodic table where elements, as we understand them, simply don't exist?

'Practically speaking?' he sighs, thinking of the potential arguments among chemists and physicists. 'All hell would break loose.'

★ ★ ★

This theoretical work is more than just a thought experiment. It's explaining the fundamental rules of our universe. These, in turn, help us to make breakthroughs in astrophysics, computing, nanomachines, energy and medicine. But the only way we can really show what's going on is by getting there. Making these elements. And that's where Yuri Oganessian and the other element hunters come in.

It's why I've come back to Dubna. Back, one last time, to JINR. Oganessian has no plan to stop making elements. As with Seaborg, Flerov and Ghiorso, there is something that drives him to still pick at the edges of the periodic table into his mid-eighties. Oganessian's search takes many forms. He's still obtaining meteorites to check for traces of superheavy elements, hoping to find the telltale marks in olivine crystals that prove an impact from something heavier than uranium. 'They're like a flying lab,' he muses. 'For detecting superheavies, they are ideal.' Now the superheavy elements are better understood, he's also able to make better predictions about where they may exist (or may have existed) in nature. Looking in hot springs, it turns out, was all wrong – if their traces are anywhere on Earth, it will likely be at the poles, falling from the sky as cosmic rays that became trapped in ice. But Oganessian's real weapon is a short walk from his office, down the snowy boulevards of Dubna.

'Titanium-50 has two more protons, but the same quantity of neutrons, as calcium-48,' Sergei Dmitriev explains, outlining the Russian plan. 'The stability of the compound nucleus will be lower, and the cross section decreases at least one order of magnitude. When we produce element 118 [at the moment], we produce about one nucleus a month. If it's an order of magnitude less, it would be one nucleus every 10 months. We're not so rich as to be able to spend that much time with a cyclotron.' That's why, while RIKEN have chosen to upgrade their equipment to hunt for elements 119 and 120, the JINR team have decided to build an entirely new cyclotron. We're standing in the room I last saw on a monitor in Dmitriev's office. Here, standing in the room as machine parts are pressure-hosed down with acetone, I can appreciate what this will become: the DC-280 cyclotron. The Russians call it the Superheavy Element Factory. When it comes online, it will be the most formidable element-hunting machine the world has ever seen.

Oganessian walks up to it, places his hand on the metal and gives it an affectionate pat. His new toy has had quite the journey. Parts have come from Russia, the US, the Czech Republic, Bulgaria, Romania and Slovakia. The main magnet came from Ukraine, just as the eastern fringes of the country became embroiled in a civil war: the 1,100t prize had to be transported on the back of a train through the conflict zone to make it to JINR. Nikolay Aksenov tells me that he remembers phoning up the Ukrainian factory to check the progress and hearing gunfire down the end of the line. He can laugh about it today; neither he nor the factory staff were laughing at the time.

'This is like Pandora's box,' Oganessian says. 'A new facility. A new accelerator. It's 10 times more powerful than the existing [machine]. The intensity will be 10 times more. Another technical achievement is the energy variation. I mean … I designed it for superheavy element research.'

As with any new piece of scientific equipment, the cyclotron will be tested first, going over known nuclides and explored decay chains to make sure it works. Then, gradually, it will ramp up the quantity of production. 'The first step will be a factor of 10,' Oganessian says above the pressure hose. 'Then a factor of 100. So, if the species [elements we are making] is 114 or 115, and we're producing 1 atom a day at the moment, this will produce 100 atoms a day. It's not much, but there's some mass. We can really check and find out the properties. With the new set-up, we can do so much more!'

The new machine is why the Russians have so much faith in their titanium-50 beam. Even with the drop-off in cross sections compared with calcium-48, you'll still be able to increase your chance of finding a new element by a factor of 10. It is arguably the world's best hope for making elements 119 and 120. It's also far more. If (or when) the Superheavy Element Factory works, the elements will finally be produced on a scale that will allow proper chemistry. Can you truly call an element that only exists for atoms at a time, that vanishes in less than a thousandth of a second, something you've discovered? The Superheavy Element Factory will end those debates.

In the next 10 years a whole new collection of elements will be open to investigation. Who knows what we will find. Perhaps the superheavy elements will find their place in the world; perhaps they will just help us understand it. Either way, they will stop being just 'that weird bit at the bottom of the periodic table'.

I turn to Oganessian. Despite his discoveries, despite the advances the heaviest elements have brought in decoding our universe, there's still the greatest question of all. After 60 years in superheavy research, what drives him to continue?

Oganessian shrugs. 'If you have a device that can do this, why not?'

Epilogue

The Royal Society sits at the heart of London, UK, a stone's throw from Buckingham Palace. It's one of the most prestigious scientific institutions in the world. Its Charter Book, containing the signatures of every fellow since its foundation in 1660, is a formidable list of names: Isaac Newton, Michael Faraday, Stephen Hawking. The atomic and elemental discoverers also grace its rolls. Look through its pages and you'll find the signatures of Ernest Rutherford, Enrico Fermi, Lise Meitner and Glenn Seaborg (women weren't allowed to be made fellows until 1945, so Marie Curie isn't present). The Society's main reception halls are upstairs; here, the greatest minds of each generation gather to pay their respects to the giants of their age.

Tonight, the room has filled to honour Yuri Oganessian.* It's an inauspicious time for such a gathering: only a week earlier, former spy Sergei Skripal and his daughter Yulia were poisoned in Salisbury. The international community has pointed the finger squarely at the Russian government, and the UK Foreign Office has been advised to keep the meeting as low-key as possible. Trouble is also brewing in the superheavy community, this time over whether there was enough evidence to confirm that the Dubna–Livermore team discovered element 117. Nobody really doubts that they found tennessine; it's an argument about the semantics of what counts when it comes to element discovery. Even so, bad blood stirs once again. When the Russian team were confronted with the doubts at a conference, they rose in unison and walked out.

* Oganessian was being awarded an honorary fellowship from the Royal Society of Chemistry as part of the UK–Russia Year of Science and Education.

But this isn't the time. It's a moment to breathe, reflect and celebrate.

The element hunters' creations shaped our world. Their names have become legend. It's time for a Charter Book of their own.

Edwin McMillan went on to lead Berkeley Lab until 1974. In 1984 he suffered the first of a series of strokes and eventually passed away from complications of diabetes in 1991, aged 83. You can see his Nobel Prize medal at the National Museum of American History in Washington, DC.

Phil Abelson became interested in naval power. In 1946 he published a report advocating for nuclear-powered submarines – now a standard branch of navies across the world. He died in 2004.

The leading element discoverer of all time remains Al Ghiorso. Never willing to sit back and enjoy retirement, Ghiorso continued to work at Berkeley Lab into old age. The last of the original element hunters, he died in 2010, aged 95. His wife Wilma, who had conspired with Helen Seaborg to get Al on the team, died in 1995.

Kenneth Street returned to Berkeley and eventually became the lab's deputy director. He died in 2006.

Gregory Choppin went on to teach at Florida State University, where the chemistry professorship is named in his honour. He died in 2015.

Bernard Harvey had a long and successful career at Berkeley. He died in 2016.

James Harris retired in 1988. A tireless advocate for scientists of colour, he also became a champion for access to education in underprivileged communities. A father of five, Harris died in 2000.

Ken Hulet retired for personal reasons shortly after the Dubna–Livermore collaboration began. He died in 2010.

Many of the element makers are still alive today. Matti Nurmia and Matti Leino continue to teach at the University of Jyväskylä in their native Finland, where Kari and Pirkko Eskola still reside.

GSI's Peter Armbruster enjoys happy retirement in France.

Gottfried Münzenberg and Sigurd Hofmann are largely retired, but both regularly attend superheavy element conferences.

Dawn Shaughnessy is still at Livermore, a dedicated advocate for both women in science and galaxies far, far away. In 2018 she was made a fellow of the American Chemical Society. Nancy and Mark Stoyer are both still there too, living proof that chemists and physicists can get along.

James Roberto and Kevin Smith have retired from Oak Ridge, but the rest of the team keep pulling small miracles from their atomic forge.

All of the Russian and Japanese teams are still involved in the hunt for new elements. They search. They hope. They dream.

Victor Ninov, the scientist alleged to have faked data for the discovery of element 118, never returned to the superheavy community and is no longer in contact with his former friends. He lives in California.

Several personalities who touched the superheavy world achieved their greatest successes away from element discovery. Three of them won the Nobel Prize for their work.

Emilio Segrè used the Berkeley bevatron to discover the antiproton – the proton's antimatter counterpart. He died in 1989.

Luis Alvarez won the prize for his contributions to elementary particle physics, and is widely regarded as one of the greatest scientists of the modern age. In later life, he developed the Alvarez hypothesis: that the dinosaurs were wiped out as the result of an asteroid impact. He died in 1988.

Finally, Melvin Calvin turned his focus to plant biology. Applying his chemical knowledge to photosynthesis, he mapped out the Calvin cycle – the reactions essential to life on Earth. He died in 1997.

The scientists who paved the way for the element makers have never been forgotten either.

Marie-Anne Paulze Lavoisier survived the French Revolution and went on to marry the British physicist Count

Rumford. She kept her first husband's last name as a mark of her devotion to him until her death in 1836.

Ernest Rutherford, largely regarded as one of the greatest scientists of all time, died in 1937. He is interred in Westminster Abbey.

Frederick Soddy, the co-discoverer of transmutation with Rutherford, had a less dignified end. Although he would go on to win the Nobel Prize for his work on isotopes, in the 1920s he developed controversial ideas about economics and anti-Semitic views. He died in 1956.

James Chadwick was knighted for his work on the Manhattan Project, and later became master of Gonville and Caius College at the University of Cambridge. He died in 1974.

Otto Hahn became one of the most influential figures in the newly formed West Germany. He is seen by many as the model of scientific integrity. Haunted that his discovery of atomic fission had caused nuclear weapons, he became a major advocate for nuclear disarmament.

Lise Meitner was named Woman of the Year by the US National Press Club in 1946 and is arguably the most influential woman scientist since Marie Curie. She remained lifelong friends with Hahn. Both died in 1968.

Laura Fermi went on to write a biography of her husband's work. Although countless volumes have since been written about 'The Pope', hers remains the most intimate and best. She died in 1977, survived by their two children.

Helen Seaborg went on to have seven children with Glenn and is remembered today as a child welfare advocate. She spent so much time walking with her husband that she developed hiking routes across California. Today, you can follow in the Seaborgs' footsteps as part of the American Discovery Trail. She died in 2006.

Kenneth Bainbridge returned to Harvard University after the Manhattan Project and later became head of the university's physics department. His experiences with nuclear power convinced him to dedicate the rest of his life to nuclear disarmament. He died in 1996, aged 91.

Maria Goeppert Mayer died in 1972. Her nuclear shell structure model is still the key to much of current superheavy element research. The Goeppert Mayer crater on Venus is named in her honour.

Jimmy Robinson's daughter, Becky Miller, works to support veterans of the US atomic programme in Florida. Through her, the Robinson contribution to science lives on: Miller's daughter majored in chemistry.

Ken Gregorich retired from Berkeley in 2018, but Jacklyn Gates continues work at the cyclotron.

Walter Loveland stays involved in the community at Oregon State.

Paul Karol lectures at Carnegie Mellon University and is now a key member of the IUPAC/IUPAP joint working party that decides when an element has been discovered.

David Hinde and Nanda Dasgupta are pushing boundaries at ANU (and yes – they would still like you to fill their fridge with beer if you pop by).

Heinz Gäggeler and Robert Eichler both continue their research in Switzerland.

Darleane Hoffman, now in her nineties, lives in California. She is admired among the chemistry community to the point of reverence. In 2017 *Chemical & Engineering News* voted her one of 13 women chemists who should have won the Nobel Prize. She never found an element, but given the fondness with which her colleagues remember her, perhaps she found something more.

The names mentioned are only a snapshot of 70 years of discovery. Countless researchers, theoreticians, experimentalists, technicians, professors and students from around the world have contributed days, months and years of research into the superheavy elements. New players are also emerging from China and France, eager to claim an element of their own. Their contributions are not forgotten.

In the next five years, the superheavy community's broad goals are simple. First, they hope to discover elements 119 and 120. This is a straight race between Dubna and RIKEN, and nobody knows who will come out on top. During my travels

I was told several possible names for the new elements by their potential discoverers; these will remain a secret.

The next aim is to edge ever closer to the island of stability; if we can reach it, the superheavy elements will stop being fleeting matter only glimpsed in a laboratory and can become an essential part of our world. Nobody really knows how important this will be.

Last, the element makers want to mass-produce the superheavy elements. This will allow for bigger, bolder chemistry experiments not so dependent on time. With them, we will learn more about our world. Perhaps oganesson is the end of where the periodic table still matters; perhaps it isn't. We won't know until we look.

There are dangers. The superheavy community is ageing, with not enough young blood coming through to continue the work. Funding is slipping away. And vital set-ups, such as Oak Ridge's HFIR, are under threat – the US government has, at present, no plan to replace it as it nears the end of its lifespan. Everyone believes we will have two new elements in the next five years; few are willing to be so bold about the five after that.

Back to the Royal Society. I glance over at Oganessian. Now in his late eighties, he is still the leading name in superheavy elements – the rock star physicist, the man who, more than any other, completed the seventh row of the periodic table. Although Nobel Prize deliberations are supposed to be secret, I know he has been nominated multiple times. His namesake may stay for only the blink of an eye, but his legacy will last forever.

At the start of this journey, I said that most scientists often view the final 26 elements as irrelevant. Some even question whether the superheavies, single atoms so unstable they can vanish in less than a second, are 'real' elements at all. They have no use. You can't hold a superheavy element in your hand. Chances are, as you read this, many of these elements do not exist anywhere in the universe. They are chemical unicorns.

But they are unicorns we know exist. Something, at the point where science meets the soul, drives people to explore the unknown. It's how we find the answers to questions we haven't thought of yet. The search for superheavy elements is the perfect example of this thirst for knowledge.

I can't help but feel optimistic. The heavy element community weathered the greatest storms of the twentieth century and kept on building the jigsaw of our world. Finally united, it has never been stronger.

This doesn't feel like the end of the superheavy story.

It feels like the start.

References

Alvarez, L. (1987). *Alvarez: Adventures of a Physicist.* New York: Basic Books

Armbruster, P. & Münzenberg, G. (2012). An Experimental Paradigm Opening the World of Superheavy Elements. *European Physical Journal H* 37: 237–309. DOI: 10.1140/epjh/e2012-20046-7

Atterling, H. *et al.* (1954). Element 100 Produced by Means of Cyclotron-Accelerated Oxygen Ions. *Physical Review* 95: 585–586. DOI: 10.1103/PhysRev.95.585.2

Bainbridge, K. (1975). A Foul and Awesome Display. *Bulletin of the Atomic Scientists* 31 (5): 40–46. DOI: 10.1080/00963402.1975.11458241

Barber, R. *et al.* (1993). Discovery of the Transfermium Elements. Part II: Introduction to Discovery Profiles. Part III: Discovery Profiles of the Transfermium Elements. *Pure and Applied Chemistry* 65: 1757–1814. DOI: 10.1351/pac199365081757

Carlson, P. (ed.) (1989). *Fysik I Frescati 1937–1987.* Stockholm: Gotab

Carnall, W. & Fried, S. (1976). *Proc. Symp. Commemorating the 25th Anniversary of Elements 97 and 98*, LBL-Report 4366. Berkeley: Lawrence Berkeley Laboratory

Chapman, K. (2016). What It Takes to Make a New Element. *Chemistry World.* Available from: https://www.chemistryworld.com/1017677.article

Chiera, N. *et al.* (2017). Attempt to Investigate the Adsorption of Cn and Fl on Se surfaces. *ResearchGate.* DOI: 10.13140/RG.2.2.13335.57766

Choppin, G. (2003). Mendelevium. *Chemical & Engineering News.* Available at: pubs.acs.org/cen/80th/mendelevium.html

Cochran, T., Norris, R. & Bukharin O. (1995). *Making the Russian Bomb: From Stalin to Yeltsin.* Boulder: Westview Press

Discovery of Mendelivium [sic] (1955 [film]). San Francisco: KQED

Edelstein, N. (ed.) (1982). *Actinides in Perspective: Proceedings of the Actinides – 1981 Conference.* Oxford: Pergamon

Fermi, L. (1954). *Atoms in the Family: My Life with Enrico Fermi.* Chicago: University of Chicago Press

Fields, P. *et al.* (1957). Production of the New Element 102. *Physical Review* 107: 1460–1462. DOI: 10.1103/PhysRev.107.1460

Flerov, G. & Petrjak, K. (1940). Spontaneous Fission of Uranium. *Physical Review* 58: 89. DOI: 10.1103/PhysRev.58.89.2

Garden, N. & Dailey, C. (1959). *High-Level Spill at the HILAC.* Berkeley: University of California

Ghiorso, A. to Fermi, L. (1955). Private correspondence, April

Ghiorso, A. *et al.* (1958). Attempts to Confirm the Existence of the 10-Minute Isotope of 102. *Physical Review Letters* 1: 18–21. DOI: 10.1103/PhysRevLett.1.17

Ghiorso, A. *et al.* (1993). Responses on 'Discovery of the Transfermium Elements' by Lawrence Berkeley Laboratory, California; Joint Institute for Nuclear Research, Dubna; and Gesellschaft fur Schwerionenforschung, Darmstadt Followed by Reply to Responses by the Transfermium Working Group. *Pure and Applied Chemistry* 65: 1815–1824. DOI: 10.1351/pac199365081815

Gilchriese, M. *et al.* (2002). Report from the Committee on the Formal Investigation of Alleged Scientific Misconduct by LBNL Staff Scientist Dr Victor Ninov. Lawrence Berkeley National Laboratory, March 27

Goeppert Mayer, M. (1949). On Closed Shells in Nuclei. II. *Physical Review* 75: 1969. DOI: 10.1103/PhysRev.75.1969

Goro, F. (1946). Plutonium Laboratory. *Life*, 8 July: 69–83

Harvey, B. *et al.* (1954). Further Production of Transcurium Nuclides by Neutron Irradiation. *Physical Review* 93: 1129. DOI: 10.1103/PhysRev.93.1129

Haxel, O., Jensen, J. & Suess, H. (1949). On the 'Magic Numbers' in Nuclear Structure. *Physical Review* 75: 1766. DOI: 10.1103/PhysRev.75.1766.2

Hinde, D. (2018). Fusion and Quasifission in Superheavy Element Synthesis. *Nuclear Physics News* 28: 15–22

Hoffman, D. *et al.* (1971). Detection of Plutonium-244 in Nature. *Nature* 234: 132–134. DOI: 10.1038/234132a0

Hoffman, D., Ghiorso, A. & Seaborg, G. (2000). *The Transuranium People: The Inside Story.* London: Imperial College Press

Hofmann, S. (2002). *On Beyond Uranium: Journey to the End of the Periodic Table.* London: Taylor & Francis

Hofmann, S. & Münzenberg, G. (2000). The Discovery of the Heaviest Elements. *Review of Modern Physics* 72: 733. DOI: 10.1103/RevModPhys.72.733

Holden, N. & Coplen, T. (2004). The Periodic Table of Elements. *Chemistry International* 26 (1): 8–9

Holloway, D. (1994). *Stalin and the Bomb: The Soviet Union and Atomic Energy 1939–1956*. New Haven: Yale University Press

Ikeda, N. (2011). The Discoveries of Uranium 237 and Symmetric Fission – From the Archival Papers of Nishina and Kimura. *Proceedings of the Japan Academy, Series B, Physical and Biological Sciences* 87: 371–376

Ito, K. (2002). Values of 'Pure Science': Nishina Yoshino's Wartime Discourse between Nationalism and Physics, 1940–1945. *Historical Studies in the Physical and Biological Sciences* 33: 61–86. DOI: 10.1525/hsps.2002.33.1.61

Jeannin, Y. & Holden, N. (1985). The Nomenclature of the Heavy Elements. *Nature* 313: 744. DOI: 10.1038/313744b0

Jerabek, P. *et al.* (2018). Electron and Nucleon Localization Functions of Oganesson: Approaching the Thomas-Fermi Limit. *Physical Review Letters* 120:053001.DOI:10.1103/PhysRevLett.120.053001

Johnson, G. (2002). At Lawrence Berkeley, Physicists Say a Colleague Took Them for a Ride. *New York Times*, October 2015

Joint Institute for Nuclear Research (2008). *Academician Yuri Tsolakovich Oganessian: 75th Anniversary*. Dubna: JINR

Joint Institute for Nuclear Research. (2018). *FLNR History: G. N. Flerov*. Available from: flerovlab.jinr.ru/flnr/history/flerov_cont.html

Karol, P. (1996). *On Naming the Transfermium Elements*, White Paper

Karol, P. *et al.* (2016a). Discovery of the Elements with Atomic Numbers Z= 113, 115 and 117 (IUPAC Technical Report). *Pure and Applied Chemistry* 88: 139–153. DOI: 10.1515/pac-2015-0502

Karol, P. *et al.* (2016b). Discovery of the Element with Atomic Number Z=118 Completing the 7th Row of the Periodic Table (IUPAC Technical Report). *Pure and Applied Chemistry* 88: 155–160. DOI: 10.1515/pac-2015-0501

Khariton, Y. *et al.* (1993). The Khariton Version. *Bulletin of the Atomic Scientists* 49 (4), 20–32. DOI: 10.1080/00963402.1993.11456341

Koppenhol W. *et al.* (2016). The Four New Elements are Named. *Pure and Applied Chemistry* 88: 401

Kragh, H. (2018). *From Transuranic to Superheavy Elements: A Story of Dispute and Creation*. Switzerland: Springer International Publishing

Kramer, K. (2017). Game Over for Original Kilogram as Metric System Overhaul Looms. *Chemistry World*. Available from: https://www.chemistryworld.com/3007760.article

Lachner, J. *et al.* (2012). Attempt to Detect Primordial 244Pu on Earth. *Physical Review C* 85: 015801. DOI: 10.1103/PhysRevC.85.015801

Lansdale, J. (1948). Superman and the Atom Bomb. *Harper's Magazine*, April 1948: 355

Lee, I-Y *et al.* (2001). Independent Study of the Synthesization of Element 118 at the LBNL 88-Inch Cyclotron. Lawrence Berkeley National Laboratory, January 25

Loveland, W., Morrissey, D. & Seaborg, G. (2017) *Modern Nuclear Chemistry 2nd Edition*. Hoboken: Wiley

Magueijo, J. (2009). *A Brilliant Darkness: The Extraordinary Life and Mysterious Disappearance of Enrico Fermi*. New York: Basic Books

Maly, Ya. (1965). On the Possibility of Producing Unexcited Compound Nuclei of the Heavy Transuranic Elements. *Soviet Physics–Doklady* 10: 1153–1156

McMillan, E. & Abelson, P. (1940). Radioactive Element 93. *Physical Review* 57: 1185. DOI: 10.1103/PhysRev.57.1185.2

Medvedev, Z. (1999). Stalin and the Atomic Bomb, in K. Coates, ed., *The Short Millennium*. Nottingham: Spokesman Books, 50–65

Meitner, L. & Frisch O. (1939). Disintegration of Uranium by Neutrons: A New Type of Nuclear Reaction. *Nature* 143: 239. DOI: 10.1038/143239a0

Nazarewicz, W. (2018). The Limits of Nuclear Mass and Charge. *Nature Physics* 14: 537–541. DOI: 10.1038/s41567-018-0163-3

Ninov, V. *et al.* (1999). Observation of Superheavy Nuclei Produced in the Reaction of 86Kr with 208Pb. *Physical Review Letters* 83: 1104–1107. DOI: 10.1103/PhysRevLett.83.1104 [Retracted]

Nishina, Y. (1947). A Japanese Scientist Describes the Destruction of his Cyclotrons. *Bulletin of the Atomic Scientists* 3: 145–167. DOI: 10.1080/00963402.1947.11455874

Nobel Prize (1938). The Nobel Prize in Physics. Available from: https://www.nobelprize.org

Öhrström, L. & Holden, N. (2016). The Three-Letter Element Symbols. *Chemistry International* 38: 4–8. DOI: 10.1515/ci-2016-0204

Periodic Videos (2013). *Seaborgium*, Periodic Table of Videos. Available from: https://youtu.be/UWq0djr790E

Periodic Videos (2017). *The Element Creator*, Periodic Table of Videos. Available from: https://youtu.be/1VaY9N7Alq0

Periodic Videos (2018). *The Office of Georgy Flyorov*, Periodic Table of Videos. Available from: https://youtu.be/UMa21BUinsI

Principe, L. (2013). A Fresh Look at Alchemy. *Chemistry World*. Available from: https://.www.chemistryworld.com/6296.article

Pyykkö, P. (2016). Is the Periodic Table All Right ('PT OK')? *EPJ Web of Conferences* 131: 01001. DOI: 10.1051/epjconf/201613101001

Rhodes, R. (1987). *The Making of the Atomic Bomb*. London: Simon & Schuster

Robinson, J. (1944). Speech to Lion's Club. Memphis, US, 17 October

Sargeson, A. *et al.* (1994). Names and Symbols of Transfermium Elements. *Pure and Applied Chemistry* 66: 2419–2421

Schädel, M. & Shaughnessy, D. (eds) (2014). *The Chemistry of Superheavy Elements*. Heidelberg: Springer

Schwartz, A. & Boring, W. (2018). *Superman: The Golden Age Dailies, 1944–1947*. New York: IDW

Seaborg G. (1946). The Impact of Nuclear Chemistry. *Chemical & Engineering News* 24: 1192: 375–381

Seaborg, G. (1951). Nobel Banquet Speech. Available from: https://.www.nobelprize.org

Seaborg, G. (1978). Stanley Thompson – a Chemist's Chemist. *Chemtech* 8: 408

Seaborg, G. (1989). *Nuclear Fission and the Transuranium Elements*. Berkeley: Lawrence Berkeley Laboratory

Seaborg, G. (1996). *A Scientist Speaks Out: A Personal Perspective on Science, Society and Change*. Singapore: World Scientific

Seaborg, G. & Corliss, W. (1971). *Man and Atom*. New York: EP Dutton & Co.

Seaborg, G. & Seaborg, E. (2001). *Adventures in the Atomic Age: From Watts to Washington*. New York: Farrar, Straus and Giroux

Seaborg, G. (ed.) (1979). *Proc. Symp. Commemorating the 25th Anniversary of Elements 99 and 100*, LBL-Report 7701. Berkeley: Lawrence Berkeley Laboratory

Segrè, E. (1939). An Unsuccessful Search for Transuranic Elements. *Physical Review* 55: 1104. DOI: 10.1103/PhysRev.55.1104

Slater, J. (1973). Putting Soul into Science. *Ebony*, May: 144–150

Snow, C. (1981). *The Physicists*. Boston: Little, Brown

Superman Strip Gives Office of Censorship Atomic Headache. *Independent News*, September–October 1945

Sutton, M. (2006). Transmutations and Isotopes. *Chemistry World*. Available from: https://www.chemistryworld.com/3004868.article

The Breath of the Dragon. *Newsletter for America's Atomic Veterans*, ed. E. Ritter, October 2013: 3–11

Thompson, S. *et al.* (1950). The New Element Californium (Atomic Number 98). *Physical Review* 80: 790–796. DOI: 10.1103/PhysRev.80.790

Thompson, S., Ghiorso, A. & Seaborg, G. (1950). The New Element Berkelium (Atomic Number 97). *Physical Review* 80: 781–789. DOI: 10.1103/PhysRev.80.781

Thornton, B. & Burdette, S. (2014). Nobelium Non-Believers. *Nature Chemistry* 6: 652. DOI: 10.1038/nchem.1979

Thornton, B. & Burdette, S. (2017). Frantically Forging Fermium. *Nature Chemistry* 9 (7): 724. DOI: 10.1038/nchem.2806

Thornton, B. & Burdette, S. (2019). Neutron stardust and the elements of Earth. *Nature Chemistry* 11 (1): 4. DOI: 10.1038/s41557-018-0190-9

US Air Force (1963). *History of Air Force Atomic Cloud Sampling.* Washington DC: US Air Force

Wapstra, A. (1991). Criteria That Must Be Satisfied for the Discovery of a New Chemical Element to Be Recognized. *Pure and Applied Chemistry* 63: 879–886. DOI: 10.1351/pac199163060879

Acknowledgements

This book has been the hardest task I've ever attempted. Throughout, my only goal has been to record this amazing story truthfully, fairly and accurately – albeit while having as much fun as possible. The adventures I've experienced in my research are some of the most potent memories of my life. It has been a two-year delirium, complete with dramatic highs and painful lows. It is thanks to the following people that *Superheavy* is with you today.

First, my fact checkers. Mark and Nancy Stoyer both read the book from cover to cover in an early draft, offering helpful suggestions and kind words of encouragement. So too did Sheila Chapman (love you, Mum!); Jennifer Newton, who gave proceedings some added pep; Nessa 'Super Science Girl' Carson, who had no reason to read anything but did so anyway; Hilary Sklar, who wisely told me not to dash off in strange directions; and Allison Holloway, who is my hero. Almost everyone quoted also read several chapters and offered gentle corrections to very stupid errors – any mistakes that remain are mine alone. Special mentions must go to Hideto En'yo, Julie Ezold, Jacklyn Gates, Sigurd Hofmann, Paul Karol, Matti Leino, Gottfried Münzenberg, Matti Nurmia, Yuri Oganessian and James Roberto for their kindness, patience and wisdom, and to Becky Miller, who agreed to fact-check the story of her father Jimmy Robinson. Thanks are also due to the team at Bloomsbury Sigma for letting me tackle such an ambitious story, and the advice and support they have offered along the way.

Support comes in many shapes and sizes. Friends and family have had to put up with me talking about one subject at every possible moment; they have done so with good grace and humour. The *Chemistry World* team at the Royal Society of Chemistry have been outstanding: Adam Brownsell, Phillip Broadwith, Jamie Durrani, Katrina Kramer, Scott Ollington, Christopher Pink, Philip Robinson, Emma Stoye,

Rebecca Trager, Ben Valsler, Patrick Walter and Neil 'rackets, not squash' Withers. I have no idea how they put up with me. Mention must also go to Alex and Jen Corbett, Marco Galea, Alex Parnell and Adam Roberts, who also got regular doses of science history whether they wanted them or not.

All-access books don't get very far without access. For that, I must thank the teams at the Australian National University (ANU), the GSI Helmholtz Centre for Heavy Ion Research (GSI), the Joint Institute for Nuclear Research (JINR), Lawrence Berkeley National Laboratory, Lawrence Livermore National Laboratory, Oak Ridge National Laboratory, the Institute of Physical and Chemical Research (RIKEN) and Stockholm University, as well as the scientists from other institutions – active or retired – who gave up their time to be interviewed in person, on the phone or both. The press offices at Argonne National Laboratory and Los Alamos National Laboratory also offered information crucial to the book.

Friendships forged on travels will, I hope, last a lifetime. I'll never forget the long road to Dubna with Alexander Madumarov, raiding Gilman Hall with Lars Öhrström, almost getting kidnapped with Allana Sliwinski or sticker hunts with Shanon Smith. Stuart Cantrill helped with arrangements for my Scandinavian adventure (and gave things a final check), Alice Williamson gave me a chance to test my anecdotes in public, Kristy Turner and Sheila Kanani gave specialist input and James Holloway provided expertise on DC Comics and weird fiction.

Last, thanks to Gamera. You're a dopey tortoise who almost destroyed the first draft of this manuscript, but you're my little buddy.

Index

Periodic Table of Elements

Atomic Number → 1

H ← Symbol

Name → Hydrogen

1.008 ← Atomic Weight

1 IA	2 IIA	3 IIIB	4 IVB	5 VB	6 VIB	7 VIIB	8 VIIIB	9 VIIIB
1 **H** Hydrogen 1.008								
3 **Li** Lithium 6.94	4 **Be** Beryllium 9.0121831							
11 **Na** Sodium 22.98976928	12 **Mg** Magnesium 24.305							
19 **K** Potassium 39.0983	20 **Ca** Calcium 40.078	21 **Sc** Scandium 44.955908	22 **Ti** Titanium 47.867	23 **V** Vanadium 50.9415	24 **Cr** Chromium 51.9961	25 **Mn** Manganese 54.938044	26 **Fe** Iron 55.845	27 **Co** Cobalt 58.933194
37 **Rb** Rubidium 85.4678	38 **Sr** Strontium 87.62	39 **Y** Yttrium 88.90584	40 **Zr** Zirconium 91.224	41 **Nb** Niobium 92.90637	42 **Mo** Molybdenum 95.95	43 **Tc** Technetium (98)	44 **Ru** Ruthenium 101.07	45 **Rh** Rhodium 102.90550
55 **Cs** Caesium 132.90545196	56 **Ba** Barium 137.327	57 - 71 Lanthanoids	72 **Hf** Hafnium 178.49	73 **Ta** Tantalum 180.94788	74 **W** Tungsten 183.84	75 **Re** Rhenium 186.207	76 **Os** Osmium 190.23	77 **Ir** Iridium 192.217
87 **Fr** Francium (223)	88 **Ra** Radium (226)	89 - 103 Actinoids	104 **Rf** Rutherfordium (267)	105 **Db** Dubnium (268)	106 **Sg** Seaborgium (269)	107 **Bh** Bohrium (270)	108 **Hs** Hassium (269)	109 **Mt** Meitnerium (278)

57 **La** Lanthanum 138.90547	58 **Ce** Cerium 140.116	59 **Pr** Praseodymium 140.90766	60 **Nd** Neodymium 144.242	61 **Pm** Promethium (145)	62 **Sm** Samarium 150.36	63 **Eu** Europium 151.964	64 **Gd** Gadolinium 157.25
89 **Ac** Actinium (227)	90 **Th** Thorium 232.0377	91 **Pa** Protactinium 231.03588	92 **U** Uranium 238.02891	93 **Np** Neptunium (237)	94 **Pu** Plutonium (244)	95 **Am** Americium (243)	96 **Cm** Curium (247)